Y0-CCH-870

The Research Report Series of the Institute for Social Research is composed of significant reports published at the completion of a research project. These reports are generally prepared by the principal research investigators and are directed to selected users of this information. Research Reports are intended as technical documents that provide rapid dissemination of new knowledge resulting from ISR research.

RESEARCH REPORT SERIES, INSTITUTE FOR SOCIAL RESEARCH

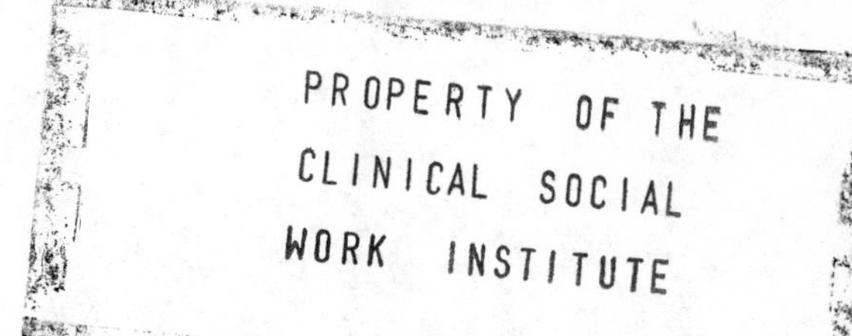

Compensating for Missing Survey Data

Graham Kalton

Survey Research Center
Institute for Social Research
The University of Michigan

1983

Income Survey Development Program, Survey Development Research Center in Nonresponse and Imputation Contract No. HEW-100-79-0127, Report on Additional Task 1.

Library of Congress Cataloging in Publication Data:

Kalton, Graham.
Compensating for missing survey data.

(Research report series / Institute for Social Research)
Bibliography: p.
1. Social surveys—Methodology. I. Title.
II. Series: Research report series (University of Michigan. Institute for Social Research)
HN29.K35 1983 301'.0723 82-12106
ISBN 0-87944-282-4 (pbk.)

ISR Code No. 9016

Published in 1983 by:
Institute for Social Research,
The University of Michigan, Ann Arbor, Michigan

6 5 4 3 2 1

Manufactured in the United States of America

TABLE OF CONTENTS

ACKNOWLEDGEMENTS

The simulation studies described in Section 4.3 were carried out with Robert L. Santos, who is a joint author of this section of the report. Robert Santos compiled the bibliography on nonresponse and missing data compensation techniques and provided a great deal of valuable assistance in the preparation of the rest of the report. His contributions are greatly appreciated. I should also like to record my thanks to Daniel Kasprzyk who provided many helpful comments on a draft version of the report. In addition I should like to thank Barbara Bailar, Brenda Cox, Brian Greenberg, James Fagan and Frederick Scheuren for drawing my attention to some inaccuracies in the draft version and making other useful comments on it.

This report was made pursuant to contract HEW-100-79-0127 with the Department of Health and Human Services. It was prepared for the Income Survey Development Program by the Survey Development Research Center in Nonresponse and Imputation which was located at the Survey Research Center, University of Michigan.

SUMMARY

This report is concerned with issues of handling missing data in surveys, with a focus on the needs of the 1978 and 1979 Research Panel surveys of the Income Survey Development Program (ISDP).

The report comprises five chapters. The first chapter discusses the types of nonresponse that give rise to missing data in surveys, in particular distinguishing between total (unit) and item nonresponse. It also outlines the bias that nonresponse can cause in survey estimates of means, totals, variances and covariances. A brief account of the total and item nonresponse encountered in the 1978 ISDP Research Panel is provided.

Chapter 2 discusses the assumptions commonly made in compensating for nonresponse, and proposes some criteria for assessing compensation procedures. These criteria include the precision of the resulting survey estimators, the ease of estimating the standard errors of these estimators, and the ability of the procedures to generate reasonable estimators for a range of population parameters. The similarities between two general classes of compensation procedures, weighting adjustments and imputation, are examined.

Weighting adjustments, which are primarily employed to compensate for total nonresponse, are reviewed in Chapter 3. The methods covered include: population weighting adjustments, weighting the sample to some external source such as the Decennial Census (as in post-stratification); sample weighting adjustments, weighting different classes of the sample by the inverses of their response rates; raking ratio estimation; and weighting according to estimated response probabilities.

Chapter 4 describes a range of imputation methods used to assign values for missing item nonresponses. The methods reviewed include deductive imputation, cold-deck and hot-deck imputation, mean-value and random imputation within classes, flexible matching imputation, distance function matching and regression imputation. The results of two simulation studies, using data from the 1978 ISDP Research Panel, are presented to illustrate and compare the properties of some of these procedures with regard to the estimation of univariate statistics. (A companion report by Santos, 1981a, employs these simulation studies to examine the effects of imputation on multivariate statistics). The chapter also discusses several issues arising with the use of imputation, including the effect of imputation on measures of the relationships between variables, the need for imputed values to satisfy edit checks, problems of handling records with several missing responses, and the choice of control variables for forming imputation classes.

The final chapter, Chapter 5, discusses the use of multiple imputations as advocated by Rubin (1978, 1979b), and the problems of applying missing data compensations to weighted data sets, where the sample units were selected with unequal probabilities. A version of a multiple imputation procedure, termed a Repeated Replication Imputation Procedure, is described in an Appendix by Kish.

A bibliography on nonresponse and nonresponse adjustments is provided at the end of the report.

1. Nonresponse in Surveys

1.1 Introduction

The purpose of this report is to review procedures for handling missing data in surveys, with a focus on the needs of the national panel surveys of the Income Survey Development Program (ISDP).

The ISDP is a program of the Department of Health and Human Services (DHSS). In collaboration with the Bureau of the Census it conducted four major field studies between October, 1977 and August, 1980 - two national household panel surveys and two studies which were confined to a number of local areas. The two national surveys, to which this report relates, are known as the 1978 and 1979 Research Panels. In both panels persons were interviewed at three month intervals; they were followed to new addresses if they moved. There were five waves of data collection in the 1978 Panel and six in the 1979 Panel. Both panels collected detailed intra-year information on household composition, money and in-kind income, assets, liabilities, program eligibility criteria and program participation, labor force participation, taxes and selected topics of policy interest. The sample size for the 1978 Panel was 2,358 households and that for the 1979 Panel about 11,300 households. Further details on the ISDP and the Research Panels are provided by Ycas and Lininger (1980).

The extensive amount and sensitivity of the data collected in the ISDP Research Panels, along with the longitudinal aspect of their designs, gave rise to a greater degree of missing data than would be encountered in a single one-time survey. In consequence, the issue of compensation for missing data is a special concern for the analysis of the Panels' data. This issue is being examined by a team from the Survey Research Center at the University of Michigan in conjunction with DHSS. This report is part of that examination.

The report is organized as follows. The next two sections of the present chapter discuss the types of nonresponse that give rise to missing data in surveys and the bias that nonresponse can cause in survey estimates. The chapter concludes with a brief account of the nonresponse encountered in the 1978 Research Panel. Chapter 2 discusses the assumptions made when compensating for nonresponse, and develops some desirable criteria for compensation procedures. It brings out the similarities and differences between two general classes of compensation procedures: weighting adjustments and imputation. Weighting adjustments, which alter the weights of respondents to compensate for nonrespondents, are then discussed in detail in Chapter 3 and imputation procedures, which assign values to missing responses, are taken up in Chapter 4. Multiple imputations and compensations when sampled units are selected with unequal probabilities are discussed in Chapter 5.

1.2 Types of Nonresponse

For subsequent discussion it is important to distinguish between three types of nonobservation: noncoverage, unit nonresponse and item nonresponse.

Noncoverage denotes a failure to include some units of the survey population in the sampling frame; in consequence, they have no chance of appearing in the sample. Although not strictly a type of nonresponse, we will loosely classify it as such for convenience of exposition throughout this report. With an area sample design, noncoverage can occur because the defined areal units fail to cover all the survey population or because the listings made within the final stage segments are incomplete.

Unit (or total) nonresponse occurs when no information is collected from a sample unit. It may be caused by a refusal, by a failure to contact the unit (not at home), by the inability of the unit to cooperate (perhaps because of illness or a language barrier), by the unit not being found (for instance, movers in a panel survey), or by completed questionnaires being lost (see Kish, 1965, Section 13.4).

Item nonresponse occurs when the unit cooperates in the survey but fails to provide answers to some of the questions. It may arise because the informant lacks the information necessary to answer the question, including his failure to make the effort required to ascertain the information by retrieving it from his memory or by consulting his records - usually described as "Don't knows"; because he refuses to give an answer, perhaps on the grounds that he finds the question sensitive, embarrassing, or considers it irrelevant to his perception of the survey's objectives; because the interviewer fails to record the answer; or because the response is subsequently rejected at an edit check on the grounds that it is inconsistent with other responses - this category may include an inconsistency arising from a coding or punching error occurring in the transfer of the response to the computer data file.

With regard to compensation for nonresponse, the importance of the distinction between noncoverage, unit and item nonresponse resides in the amount of information available about the nonrespondents. The survey itself can provide no information about units not covered by its sampling frame; if compensations are to be made for noncoverage, they must be based on some external source, such as the Decennial Census.[1] The survey can provide some information about unit nonrespondents, but the amount is usually fairly limited. At the least the sampling unit in which the nonrespondent is located will be known, together with the stratum characteristics of that sampling unit (e.g. region, urbanicity); additional information may be available from interviewers' observations (type of dwelling unit, perhaps race of head of household) and from the sampling frame (for instance, administrative information from the records of Supplemental Security Income - SSI - recipients for the SSI samples in the 1978 and 1979 Research Panels). With item nonresponse the responses to all the items that the unit has answered provide a considerable amount of information about the unit. As will be discussed in subsequent sections, the amount of information available for a particular type of nonresponse has a strong bearing on the form of compensation employed: in brief, noncoverage is commonly treated by some form of post-stratification weighting adjustment, unit nonresponse by some form of post-stratification or response rate weighting

---

[1]It should be noted that in employing an external source for making noncoverage adjustments, consideration needs to be given to the external source's own noncoverage problems.

adjustment, and item nonresponse by some form of imputation procedure.

While the distinction between unit and item nonresponse in terms of the amount of information available on the nonrespondents is a useful broad generalization, it is one of degree and not always clearcut. Sometimes a great deal may be known about unit nonrespondents from the sampling frame, in which case compensations for unit nonresponse may be better made by imputation than by weighting. On the other hand, a respondent may break off an interview at an early stage, or may refuse to answer most of the questions, in which case weighting may be more appropriate than imputation.

1.3 Nonresponse Bias

In most surveys no compensation is made for missing data. This section examines the bias resulting from confining the survey analyses to the available data.

For simplicity we will consider a simple random sample of size n drawn from a population of size N, and we will concentrate on a single variable y. With only one variable there is no need to distinguish between the types of nonresponse discussed in the previous section; units not covered by the sampling frame, unit nonrespondents and item nonrespondents will, therefore, be termed simply nonrespondents. We will further assume that, on conceptually repeated applications of the survey under the same essential survey conditions, certain units in the population always respond and the remainder never respond. Thus, the population can be thought of as comprising two strata - respondents and nonrespondents. In reality this division of the population into two strata is an oversimplification because, for some of the units at least, chance plays a part in determining whether they respond or not. The simplified model is attractive, however, because its tractability leads to some informative results. For this reason it will be employed throughout this report. The reader is referred to Platek *et al*. (1978) for the development of a probability response model for the study of the effects of nonresponse.

Let R be the number of respondents and M be the number of nonrespondents (M for missing) in the population, with $R + M = N$; the corresponding sample quantities are r and m, with $r + m = n$. Let $\bar{R} = R/N$ and $\bar{M} = M/N$ be the proportions of respondents and nonrespondents in the population and let $\bar{r} = r/n$ and $\bar{m} = m/n$ be the response and nonresponse rates in the sample. The population total and mean are given by $Y = Y_r + Y_m = R\bar{Y}_r + M\bar{Y}_m$ and $\bar{Y} = \bar{R}\bar{Y}_r + \bar{M}\bar{Y}_m$, where Y_r and $\bar{Y}_r$ are the total and mean for respondents and Y_m and $\bar{Y}_m$ are the same quantities for the nonrespondents. The corresponding sample quantities are $y = y_r + y_m = r\bar{y}_r + m\bar{y}_m$ and $\bar{y} = \bar{r}\bar{y}_r + \bar{m}\bar{y}_m$.

If no compensation is made for nonresponse, the respondent sample mean $\bar{y}_r$ is used to estimate $\bar{Y}$. Its bias is given by $B(\bar{Y}_r) = E(\bar{y}_r) - \bar{Y}$. The expectation of $\bar{y}_r$ can be obtained in two stages, first conditional on fixed r and then over different values of r, i.e. $E(\bar{y}_r) = E_1E_2(\bar{y}_r)$ where E_2 is the conditional expectation for fixed r and E_1 is expectation over different values of r. Thus

$$E(\bar{y}_r) = E_1[\Sigma E_2(y_{ri})/r] = E_1(\bar{Y}_r) = \bar{Y}_r.$$

Hence the bias of $\bar{y}_r$ is given by

$$B(\bar{y}_r) = \bar{Y}_r - \bar{Y} = \bar{M}(\bar{Y}_r - \bar{Y}_m). \quad (1.3.1)$$

Equation (1.3.1) shows that $\bar{y}_r$ is approximately unbiased for $\bar{Y}$ if either the proportion of nonrespondents $\bar{M}$ is small or the mean for nonrespondents, $\bar{Y}_m$, is close to that for respondents, $\bar{Y}_r$. Since the survey analyst usually has no direct empirical evidence on the magnitude of $(\bar{Y}_r - \bar{Y}_m)$, the only situation in which he can have confidence that the bias is small is when the nonresponse rate is low. However, in practice, even with moderate $\bar{M}$ many survey results escape sizeable biases because $(\bar{Y}_r - \bar{Y}_m)$ is fortunately often not large.

Consider now the estimation of the population total Y. With complete response the simple inflation estimator Fy, where the inflation factor $F = N/n$ is the inverse of the sampling fraction f, is an unbiased estimator of Y. The application of this inflation factor to the sample total for respondents gives an estimator $\hat{Y} = Fy_r$, with expectation

$$E(\hat{Y}) = FE_1E_2(r\bar{y}_r) = FE_1(r\bar{Y}_r) = FfR\bar{Y}_r = Y_r. \quad (1.3.2)$$

Its bias as an estimator of Y is thus

$$B(\hat{Y}) = Y_r - Y = -Y_m = -M\bar{Y}_m.$$

This bias is small only if the total for the nonrespondents is small. If the y-variable is a non-negative quantity, as is generally the case with survey variables, Y_m is small only if M is small, or the Y_{mi} values are zero or close to zero.

Next consider the estimation of the difference between two subclass means, $\bar{Y}_a - \bar{Y}_b$, with the earlier notation extended to $\bar{Y}_a = \bar{R}_a\bar{Y}_{ra} + \bar{M}_a\bar{Y}_{ma}$ and $\bar{Y}_b = \bar{R}_b\bar{Y}_{rb} + \bar{M}_b\bar{Y}_{mb}$. Then the bias of $(\bar{y}_{ra} - \bar{y}_{rb})$ is

$$\begin{aligned} B(\bar{y}_{ra} - \bar{y}_{rb}) &= B(\bar{y}_{ra}) - B(\bar{y}_{rb}) \\ &= \bar{M}_a(\bar{Y}_{ra} - \bar{Y}_{ma}) - \bar{M}_b(\bar{Y}_{rb} - \bar{Y}_{mb}). \end{aligned} \quad (1.3.3)$$

In favorable circumstances, the biases in the individual estimators $\bar{y}_{ra}$ and $\bar{y}_{rb}$ may tend to cancel. In particular, this may happen in comparing the means for two rounds of a survey when the rounds have similar response rates.

Finally, consider the effect of missing data on the estimation of variances and covariances. The expectation of the respondent sample variance $s_r^2 = \Sigma(y_{ri} - \bar{y}_r)^2/(r - 1)$ is

$$E_1E_2(s_r^2) = E_1(S_r^2) = S_r^2 \quad (1.3.4)$$

where $S_r^2 = \Sigma(Y_{ri} - \bar{Y}_r)^2/(R - 1)$. The bias of s_r^2 as an estimator of $S^2 = \Sigma(Y_i - \bar{Y})^2/(N - 1)$ is thus

$$B(s_r^2) = S_r^2 - S^2. \quad (1.3.5)$$

Now

$$\Sigma(Y_i - \bar{Y})^2 = \Sigma(Y_{ri} - \bar{Y}_r)^2 + \Sigma(Y_{mi} - \bar{Y}_m)^2 + R(\bar{Y}_r - \bar{Y})^2 + M(\bar{Y}_m - \bar{Y})^2$$

$$= (R - 1)S_r^2 + (M - 1)S_m^2 + R\bar{M}(\bar{Y}_r - \bar{Y}_m)^2.$$

Thus, approximating $(R - 1) \simeq R$, $(M - 1) \simeq M$ and $(N - 1) \simeq N$,

$$S^2 \simeq \bar{R}S_r^2 + \bar{M}S_m^2 + \bar{R}\bar{M}(\bar{Y}_r - \bar{Y}_m)^2.$$

The bias of s_r^2 is therefore

$$B(s_r^2) \simeq \bar{M}(S_r^2 - S_m^2) - \bar{R}\bar{M}(\bar{Y}_r - \bar{Y}_m)^2. \qquad (1.3.6)$$

The first term of this bias is comparable to the bias for a mean in (1.3.1). The assumption that S_r^2 and S_m^2 are similar may, perhaps, be more realistic than the assumption that $\bar{Y}_r$ and $\bar{Y}_m$ are similar. Under this assumption, the first term is negligible. The second term reflects the effect of differences in the respondent and nonrespondent means on the estimator. If respondents and nonrespondents have the same variance ($S_r^2 = S_m^2$), s_r^2 will underestimate S^2 unless $\bar{Y}_r = \bar{Y}_m$.

For the covariance another variable, x, needs to be introduced. For this case respondents will be defined as those who provide both x and y values. A "'" is added to all symbols to draw attention to the fact that they refer to this set of respondents; thus, for instance, r' is the number of sample units providing both x and y values, and $\bar{y}_r'$ is their sample mean for the y-variable. If x values are obtained for all sample units or if x is missing only for sample units for which y is missing, $r' = r$; otherwise $r' < r$. The expectation of the sample covariance $s_{rxy}' = \Sigma(x_{ri}' - \bar{x}_r')(y_{ri}' - \bar{y}_r')/(r' - 1)$ is

$$E_1E_2(s_{rxy}') = E_1(S_{rxy}') = S_{rxy}' \qquad (1.3.7)$$

where $$S_{rxy}' = \Sigma(X_{ri}' - \bar{X}_r')(Y_{ri}' - \bar{Y}_r')/(R' - 1).$$

Using the same approach as for the variance,

$$S_{xy} \simeq \bar{R}'S'_{rxy} + \bar{M}'S'_{mxy} + \bar{R}'\bar{M}'(\bar{X}'_r - \bar{X}'_m)(\bar{Y}'_r - \bar{Y}'_m).$$

Thus the bias of s'_{rxy} as an estimator of s_{xy} is

$$B(s'_{rxy}) \simeq \bar{M}'(S'_{rxy} - S'_{mxy}) - \bar{R}'\bar{M}'(\bar{X}'_r - \bar{X}'_m)(\bar{Y}'_r - \bar{Y}'_m). \qquad (1.3.8)$$

The interpretation of this bias in the covariance follows that for the variance given above, except that s'_{rxy} does not necessarily underestimate S_{xy} if $S'_{rxy} = S'_{mxy}$. If this condition applies, the sign of $B(s'_{rxy})$ depends on the signs of $(\bar{X}'_r - \bar{X}'_m)$ and $(\bar{Y}'_r - \bar{Y}'_m)$; if these terms have the same sign, $B(s'_{rxy})$ will be negative, but if they have different signs $B(s'_{rxy})$ will be positive.

The purpose of examining the impact of nonresponse on several statistics in the above discussion is to show that nonresponse can have different effects on different statistics. In examining the properties of various compensation procedures in the following chapters, attention will need to be given to their effectiveness in counteracting these diverse effects.

1.4 Nonresponse in the 1978 ISDP Research Panel

Heeringa (1980) has carried out a detailed examination of the unit and item nonresponse encountered in the 1978 ISDP Research Panel focusing primarily on the first two waves (April and July 1978) and on missing data for the income items. Briefer accounts are provided by Kalton _et al_. (1980, 1981). A few of the findings will be summarized here in order to give an impression of the magnitude of the missing data problem, and to reveal some of its facets.

Interviews were conducted with the 1978 Research Panel on five occasions, at quarterly intervals, the first wave of interviews taking place in April 1978 and the last in April 1979. The Panel was selected in 60 Census Primary Sampling Units (PSU's) with an initial sample of 2,358 sample units, consisting of 1,947 area probability sample housing units and 411 persons drawn from a list of Supplemental Security Income (SSI) recipients. The area sample was defined as the households occupying sampled residential units and the SSI sample was defined as the households in which the sampled persons were found to be living. The initial sample of persons for the April 1978 wave of the Panel was defined as all those aged 16 and over living in the sampled households. All members of the initial sample were included in the Panel. Those who moved to new addresses were followed and interviewed; however, for cost reasons, those moving beyond fifty miles of the Panel's PSU's were not followed. Persons who moved in with members of the initial sample (or vice-versa) became part of the sample.

In discussing missing data in the 1978 Research Panel, various levels of nonresponse need to be distinguished. At any one wave of the Panel, nonresponse occurred at: (1) the household level, with no data being collected for any household member; (2) the person level, with no data being collected for one or more household members in an otherwise cooperating household; and (3) the item level, when some but not all data are missing for a sample person. The panel aspect of the design adds another dimension to the household and person levels of nonresponse: a household or person may provide data for all waves of the Panel, may provide data for some waves but not for others, or may fail to provide data for any wave.

The household level nonresponse at the first (April) wave of the Panel was 6.5%, three quarters of which were refusals. Some 2.5% to 3.0% of households were lost subsequently at each successive quarter through adults moving beyond fifty miles of the Panel's PSU's and through other movers who could not be traced. Among households remaining at their sample addresses or moving within the Panel's sample areas, there was an increase in nonresponse through the life of the Panel, but at a declining rate; refusals were the dominant cause of nonresponse at each wave. By the last wave of the Panel in April 1979 the nonresponse rate among these households had risen to 15.0%.

The wave to wave change in nonresponse rates is the net effect of obtaining responses in a later wave from households that failed to provide data in an earlier one together with the loss of some households cooperating in an earlier wave but not providing data in a later one. Attempts were made in the 1978 Research Panel to secure responses in a later wave from households that did not respond in an earlier one. The only evidence available to date on the effect of these attempts comes from the first two waves of the Panel: only 20% of the 140 nonresponding households in the April wave provided data for the July wave (of the 103 households refusing in April, 15 responded in July).

There was little nonresponse at the person level within cooperating households: in the April 1978 wave, no data were provided for only about 1.5% to 2.0% of adults in such households. This low level of person nonresponse was achieved through the use of a proxy interview with a knowledgeable family member when an adult was unavailable. (Data on 31.7% of sampled adults were provided by proxy informants in the April 1978 wave of the Panel.) Person level nonresponse thus occurred only when the adult refused to cooperate or when he was absent and no other family member felt able to respond on his behalf.

The level of item nonresponse varied considerably across items. Almost complete information was obtained for many of the easier items, including the basic demographics, but there was a fair amount of nonresponse for some of the more difficult income items. Thus, for example, in the April 1978 wave some 10.8% of area sample salaried respondents paid the same amount each payday failed to provide the amount of their paycheck either from their records or as an estimate, and some 9.7% of area sample respondents paid hourly failed to provide their regular hourly rate of pay. One of the highest levels of item nonresponse occurred with an item asking for interest received during the last three months from savings accounts. In the April 1978 wave 47.6% of area sample respondents with such accounts failed to provide this information. For many items the item nonresponse rate was found to be considerably higher when the data were collected from proxy informants than from self reporters.

As the preceding discussion indicates, the simple textbook dichotomy of nonresponse into unit and item nonresponse is inadequate for a household panel survey like the 1978 Research Panel, and in fact the distinction between these two types of nonresponse becomes extremely blurred with such a survey. The inclusion of all members of selected households in the sample and the panel aspect of the design both contribute to this situation. If one person in an otherwise cooperating household refuses to be interviewed, should this be viewed as unit or item nonresponse? For analyses conducted at the household level, this type of nonresponse may be treated as a series of item nonresponses in the household record. For person level analyses, however, it may be appropriate to treat the person as a unit nonrespondent (but noting that data for the household and for other household members may provide useful information about the nonrespondent). If a household - or person - fails to provide data on one wave of the panel, should this be viewed as unit or item nonresponse? For longitudinal analyses, this case may be treated as a series of item nonresponses in the household's longitudinal record, while for cross-sectional analyses it may be treated as unit nonresponse (but again noting that the household's data from other waves may provide valuable information about its missing responses).

The inclusion of all members of selected households in the sample also introduces complexity with regard to compensation for person-level nonresponse. Apart from a few cases where interviewers may have failed to determine whether selected housing units were occupied by households, the number of nonresponding households is known. In consequence, nonresponse weighting adjustments based on the inverse of household response rates can be readily implemented. The situation for person-level nonresponse is, however, complicated by the fact that the number of persons in nonresponding households may often not be obtained. As a result person-level response rates cannot be directly computed, but have to be estimated employing some assumption about the size of nonresponding households. The effectiveness of weighting adjustments based on person-level response rates thus depends on the appropriateness of this assumption, as well as that of the other assumptions underlying the procedure.

2. General Issues for Missing Data Compensation Procedures

2.1 Assumptions

Whenever a survey experiences missing data, assumptions must be made in order to produce estimates for the population of interest. This statement holds whether compensation procedures are used or not. Researchers sometimes justify making no compensation for missing data - that is simply analyzing the responses - on the grounds that they thereby avoid the subjective assumptions required in all compensation procedures, and that their results are in consequence more defensible. If valid, this justification would be a powerful one, for objectivity of survey results is a real concern, especially for surveys conducted by government agencies. However, on examination, the justification turns out to be less persuasive than it might at first appear.

To illustrate this point, consider the use of the respondent mean $\bar{y}_r$ to estimate the population mean $\bar{Y}$, as discussed in Section 1.3. The bias of $\bar{y}_r$ was shown in equation (1.3.1) to be $\bar{M}(\bar{Y}_r - \bar{Y}_m)$. If the nonresponse rate $\bar{M}$ is sizeable, the suitability of $\bar{y}_r$ as an estimator of $\bar{Y}$ thus depends on the assumption that the respondents and nonrespondents have similar means. In other words, in a survey with missing data even an estimation procedure which does not compensate for nonresponse incorporates an implicit assumption when estimating the overall population mean. This assumption may be technically avoided by claiming that the results apply only to the response stratum, but this begs the issue. Since the response stratum is not the population of interest, the user of the survey results is left to make his own rough adjustments. Lacking detailed information about the characteristics of nonrespondents, the user cannot be expected to adjust for the missing data as well as the original analyst would have done. Thus, although compensation procedures must depend on assumptions, it seems preferable to adopt them using as realistic assumptions as possible (and fully documenting the assumptions made and procedures used), rather than simply to analyze the respondent data, a procedure which itself depends on an often highly questionable assumption.

Every compensation procedure needs to make an explicit assumption about the characteristics of the nonrespondents. In estimating means or totals, the simplest assumption is that the respondents' means are equal to those of the nonrespondents ($\bar{Y}_r = \bar{Y}_m$). An alternative assumption is that the missing data are missing at random. Two processes can be envisaged for generating data missing at random. One is to postulate a single superpopulation from which the finite populations of respondents and nonrespondents are drawn at random; in this process, respondents always respond over conceptually repeated applications of the survey and nonrespondents never do. The second process is to consider every unit in the finite population as having the same chance of responding; in this process a respondent on one application of the survey may be a nonrespondent on another. (See, for instance, Platek et al., 1978, Platek and Gray, 1979, 1980).

The assumption that data are missing at random is in one sense somewhat less exacting than the assumption $\bar{Y}_r = \bar{Y}_m$. With the first process for generating data missing at random, the finite population means for respondents and nonrespondents may differ because of sampling variability in sampling from the superpopulation; however, their ξ-expectations over sampling from the superpopulation are equal, $E_\xi(\bar{Y}_r) = E_\xi(\bar{Y}_m)$. With the second process, a complete enumeration of the population would yield different means for respondents and nonrespondents, but the expectations of these means over repeated applications are equal. Providing that the (expected) population sizes for respondents (R) and nonrespondents (M) are much larger than the sample size (n), the sampling variability from the process of generating the data missing at random can be ignored, i.e. the respondent and nonrespondent population means may be treated as approximately equal.

In another sense, the assumption that data are missing at random is more stringent than the assumption $\bar{Y}_r = \bar{Y}_m$. The latter assumption specifies nothing about the distribution of the nonrespondents' y-values around the mean, whereas the missing at random assumption implies the equality in expectation of the respondents' and nonrespondents' distributions, and hence also of the variances and other parameters of the distributions.

The simple assumptions that $\bar{Y}_r = \bar{Y}_m$ or that the data are missing at random are rarely tenable in practice. Numerous surveys have provided ample evidence to demonstrate that response rates vary across population subgroups, and the survey variables are often associated with the characteristics of these subgroups. A natural extension of the simple assumption is first to divide the population into defined subgroups, and then to apply the simple assumption in one form or the other within each subgroup. This extension to an assumption of similarity of respondents and nonrespondents within subgroups is the basis of the most widely used compensation procedures. It is the underlying assumption of both the weighting adjustments discussed in Chapter 3 and the hot-deck imputations discussed in Chapter 4.

A problem arises when the researcher wants to take account of several characteristics in forming the subgroups within which he is prepared to assume the similarity of respondents and nonrespondents. In this case the subgroups formed as the cells of the crosstabulation of the characteristics may be too numerous to permit stable compensations to be made separately within each subgroup. This situation may be avoided by relaxing the assumption of similarity in subgroups to one which equates respondents and nonrespondents in terms of an additive model for the characteristics (i.e. assuming no interaction effects). A similar approach may also be used to handle characteristics which are continuous rather than classificatory variables. It may be preferable to assume a linear model for such variables rather than the alternative of categorizing them into a relatively small number of classes. Raking ratio estimation and regression imputation derive from these kinds of assumptions.

Compensations for missing data in surveys are generally based on equating the respondents and nonrespondents in some way after controlling for other variables. The specification of equality for the two groups is, however, not essential; one could instead specify the differences between them. Rubin (1977) and Ericson (1967), for instance, develop a Bayesian analysis incorporating subjective notions of differences between respondents and nonrespondents. The difficulty of specifying such differences in a survey context with numerous variables and a multitude of analyses severely limits the utility of this approach for general purpose survey work. It will therefore not be considered further in this report.

2.2 Compensation Procedures

The statistical literature on missing data problems can be broadly divided into that which deals with the estimation of specific parameters from a simple random sample with a known density and that which is of a general purpose nature. The concern of the present report is with the latter type of problem. We will therefore not review the sizeable literature on estimating specific parameters from incomplete data; however, some references are provided in the bibliography at the end of the report (for instance, Afifi and Elashoff, 1966, 1967, 1969; Anderson, 1957; Beale and Little, 1975; Dempster et al., 1977; Woodbury and Siler, 1966).

The compensation procedures with which we are concerned are based on the general purpose philosophy of aiming to provide reasonable estimates for many forms of analysis rather than the best estimates for particular parameters. A strong case can be made for adopting this approach in survey work. In particular, given the multitude of analyses to which a survey data set is subjected by many different researchers of varying degrees of statistical sophistication, it is unrealistic to believe that separate efficient compensation procedures could be developed for each analysis; and, even if they could, it would almost certainly be uneconomic to do so. The use of different compensation procedures may also lead to inconsistencies in the results obtained.

Although the literature on compensating for missing data in surveys is mostly concerned with general purpose techniques, the theoretical results available mainly relate to the impact of the procedures on simple descriptive statistics like means and totals. This emphasis will be reflected in this report but the general purpose nature of the compensation procedures should be constantly borne in mind. The companion report by Santos (1981a) considers the effects of compensation procedures on more complex statistics, especially those obtained from regression analyses.

General purpose compensation procedures can be separated into two types, weighting adjustments and imputation techniques. In practice these two types of procedures are often used in combination to handle different kinds of missing data. Weighting adjustments increase the weights of respondents in various subgroups of the sample in order to compensate for the subgroups' differing response rates, while imputation techniques insert values for missing responses. The dominant factor determining the choice between these two types of procedures for handling a particular kind of missing data is the amount of information available on the units involved, often broadly reflected by whether the missing data arose from noncoverage, unit response, or item nonresponse. When missing data compensations are made, noncoverage and unit nonresponse are generally treated by weighting adjustments while item nonresponse is treated by imputation.

The rationale for this situation is as follows. When little information on the unit with missing data is available - as with unit nonresponse - all the information can be used in forming the subgroups for which the weighting adjustments are made. In essence, units with missing data are matched with sets of respondents with the same characteristics on the available data. Making subgroup weighting adjustments to the respondents to compensate for the nonrespondents thus retains all the known information about the nonrespondents. The procedure of matching respondents and nonrespondents on the data available for nonrespondents becomes, however, more demanding the more that is known about the nonrespondents. At some point, it will become impossible to find even one respondent who corresponds exactly with a nonrespondent in terms of the data available on the latter (this typically applies with item nonresponse). Two alternative approaches are possible when this occurs: either to discard some of the less important data about the nonrespondent until one or more matching respondents can be found, and use a weighting adjustment; or to use an imputation procedure. If only a minimal amount of uninformative data needs to be discarded, it may be appropriate to use a weighting adjustment (as, for instance, when a respondent answers only one or two of the initial questions on a questionnaire). However, when the loss of information incurred from discarding data is substantial, as it often would be with item nonresponse, the alternative approach of imputation is preferable. As discussed in the next section, imputation has a number of disadvantages compared with weighting adjustments, but it does routinely retain all the available information on units with missing data.

While the distinction between weighting adjustments and imputation is used as the major classification for the organization of this report, it should however not be assumed that the procedures are unrelated. In fact, they both operate in the same basic way of building up the respondents' data to represent the missing data. Indeed there exists an equivalent simple imputation scheme for any weighting adjustment that employs integer weights. Consider, for instance, a weighting scheme in which a respondent is chosen at random within a subgroup and is given additional weight to represent a nonrespondent from that subgroup. Providing the subgroup characteristics reflect all the information known about the nonrespondent, this scheme is equivalent to an imputation scheme in which a nonrespondent is assigned all the responses from a respondent chosen at random in the same subgroup. The two schemes will produce the same estimates from the resultant data sets. The difference between them is simply that with the weighting adjustment the records for the respondent and the nonrespondent are merged into a single record with increased weight, whereas with the imputation procedure the two records, although identical, remain separate.

An equivalence can also be established in the more typical situation when noninteger weighting adjustments are used. Suppose, for example, that in an equal probability sample the r respondents in a subgroup are each assigned the weight n/r to represent the $m = (n - r)$ nonrespondents. An equivalent imputation procedure would be to split each nonrespondent record into r parts, each with a weight of $1/r$, and assign the records from each respondent in turn to each of the r parts. The complete data set generated from this procedure would have (r + mr) records, r respondent records with unit weights and mr nonrespondent records with weights of 1/r each. Analysis of this data set would produce the same estimates as those obtained from the data set with nonresponse weighting adjustments. In certain circumstances, there exist variants of the preceding imputation scheme which are also equivalent to the weighting adjustment. If, say, r and m have a common factor r = kr' and m = km', where k, r' and m' are integers, the respondents and nonrespondents in the subgroup could be divided into k groups each of size (r' + m'), and then the preceding procedure could be applied separately within each group; that is, within each group, the nonrespondents' records are divided into r' parts each with a weight of 1/r', and each part is then assigned a respondent record from one of the r' respondents in that group. This procedure generates a data set with (r + mr') records, a smaller number than the (r + mr) in the basic version.

The similarities between weighting adjustments and imputation procedures are worth keeping in mind for they may suggest ways of devising and treating imputation procedures that overcome some of the current disadvantages to imputation as a method of compensating for missing data.

2.3 Three General Criteria in the Choice of Compensation Procedures

The previous section has discussed the basic issue of when weighting adjustments and when imputation methods should be used to compensate for missing data. This section introduces three other considerations involved in making an appropriate choice of compensation procedure. These considerations, which apply whether weighting adjustments or imputation techniques are used, are concerned with: (1) the precision of the resulting estimates; (2) the estimation of standard errors of the estimates; and (3) the suitability of the compensated data set for producing estimates for a variety of different parameters. For simplicity, the section will consider only the simplest type of compensation procedure, one applied to the sample taken as a whole. The extension to the more general situation where compensations are carried out separately for different subgroups will be taken up in the following chapters.

Throughout this chapter it is assumed that the sample of n units is drawn from the population by simple random sampling.

2.3.1 Precision of Estimators

For simplicity we will consider in this section only the estimation of the population total Y from incomplete data. As shown in Section 1.3, the simple inflation estimator $\hat{Y} = (N/n)y_r = Fy_r$ has a negative bias of $M\bar{Y}_m$ as an estimator of Y. Its use is tantamount to assuming that $\bar{Y}_m = 0$, usually an unrealistic assumption. In most cases, a more appropriate assumption would be either that $\bar{Y}_m = \bar{Y}_r$ or that the missing data are missing at random, as discussed in Section 2.1. There are a variety of methods in which compensations for the missing data can be carried out in line with these assumptions, both by weighting adjustments and imputation. For the present discussion these methods may be usefully classified into those which impose the condition $\bar{y}_m = \bar{y}_r$ and those which relax this condition to one in which the expected value of $\bar{y}_m$ over repeated applications of the method, $E_2(\bar{y}_m)$, is equal to $\bar{y}_r$.

The well-known estimator $\hat{Y}_1 = N\bar{y}_r$ represents the obvious form of weighting adjustment in which $\bar{y}_r = \bar{y}_m$. This condition can be clearly seen if $\hat{Y}_1$ is expressed as $R\bar{y}_r + M\bar{y}_r$ as an estimator of $R\bar{Y}_r + M\bar{Y}_m$. It can be expressed in the form of a simple inflation estimator using a weighting adjustment as $\hat{Y}_1 = Fy_w$ where $y_w = \Sigma w_i y_{ri}$ and $w_i = n/r$. This adjustment is the simplest form of weighting adjustment in which all respondents are assigned the same weight. In passing, it may be noted that when applied in estimating means this adjustment has no effect, yielding the same estimate as obtained without any compensation. This is readily seen from

$$\bar{y}_w = \Sigma w_i y_{ri}/\Sigma w_i = [(n/r)\Sigma y_{ri}]/[(n/r)r] = \bar{y}_r.$$

It is possible to conceive of alternative weighting adjustment schemes satisfying the condition $\bar{y}_r = \bar{y}_m$, assigning larger weights to a subset of respondents and perhaps different weights to different respondents, but such schemes may be difficult to operationalize.

A straightforward imputation procedure satisfying the condition $\bar{y}_r = \bar{y}_m$ is to impute values of $\bar{y}_r$ for all the nonrespondents. More generally, however, the imputed values could be any set of numbers that sum to $m\bar{y}_r$.

The relaxation of the condition from $\bar{y}_m = \bar{y}_r$ to $E_2(\bar{y}_m) = \bar{y}_r$ introduces a degree of flexibility which permits a wide range of alternative compensation procedures. The condition may be satisfied by a variety of weighting adjustment schemes in which weights are assigned to an equal probability (epsem) sample of respondents to compensate for the nonrespondents. There are two main aspects to the choice of weighting scheme, the sample design to be used and the size of sample to be selected. Assuming $r > m$, one obvious choice of sample size would be to take a sample of size m, and increase the weight of each selection by 1 to compensate for a nonrespondent. The sample could, for instance, be selected: by unrestricted sampling (simple random sampling <u>with</u> replacement); by SRS (simple random sampling <u>without</u> replacement); by proportionate stratified sampling with the respondents stratified in some way, perhaps by their y-values; or by systematic sampling from a list of the respondents ordered in some way. Another choice for the sample size might be cm, with the weight of each selection being increased by 1/c; any epsem scheme could then be used to select the sample. Note that the simple inflation estimator $\hat{Y}_1$ comes from the special case of this scheme when $cm = r$ and sampling is without replacement.

With $r > m$, an imputation scheme satisfying $E_2(\bar{y}_m) = \bar{y}_r$ can be obtained by selecting any epsem sample of m respondents, and assigning their values to the nonrespondents. The possible sampling schemes include any of those mentioned above. Sampling can be carried out with or without replacement, and techniques like stratification and systematic sampling can be used.

The case of $m > r$ implies that over 50% of the data on the variable under study are missing. This is seldom likely to occur in practice for the total sample and, if it were to occur, it would raise serious doubts about whether the data were usable. Nonresponse adjustments are, however, made within subgroups of the sample; the case $m > r$ may well occur in some of the subgroups and for this reason it deserves attention.

When $m > r$, an epsem without replacement sampling scheme for choosing respondents to be assigned increased weight or choosing respondents as donors of imputed values cannot be directly implemented. To cope with this problem, we generalize the without replacement sampling schemes as follows. Let $m = kr + t$, where k is a non-negative integer and $0 \leq t < r$; when $r > m$, $k = 0$ and $m = t$. Then, with weighting adjustments, t respondents are selected by the epsem sampling scheme to be given weights of $(k + 2)$ and the remaining $(r - t)$ respondents are given weights of $(k + 1)$. With imputation, all respondents serve as donors on k occasions, with an epsem sample of t of them serving as donors on one more occasion. The versions of these schemes in which SRS is used will be referred to as the SRS weighting and imputation schemes.

In this section we will derive and compare variances of the estimators $\hat{Y}$ under different compensation procedures. In these derivations, the numbers of respondents and nonrespondents in the population, R and M, will be treated as fixed. This procedure is appropriate if respondents are viewed as always responding and nonrespondents are viewed as never responding over repeated applications of the survey. If, instead, all units in the population are viewed as having the same response probability, the variances obtained are conditional ones, conditional on a given set of R respondents and M nonrespondents. All the variances are derived conditional also on a fixed number, r, of sampled respondents.

The criterion of precision of the estimator $\hat{Y}$ favors compensation procedures that require $\bar{y}_m = \bar{y}_r$ over those that require only that $E_2(\bar{y}_m) = \bar{y}_r$. To see this, note that the estimator obtained from any compensation procedure that requires $\bar{y}_m = \bar{y}_r$ can be shown to be equal to $\hat{Y}_1 = N\bar{y}_r$, and that the conditional expectation of the estimator obtained from any compensation that requires $E_2(\bar{y}_m) = \bar{y}_r$ can be shown to be equal to $\hat{Y}_1$. The variance of any $\hat{Y}$ can be expressed as

$$V(\hat{Y}) = E_1V_2(\hat{Y}) + V_1E_2(\hat{Y}) \qquad (2.3.1.1)$$

where E_2 and V_2 are conditional expectation and variance over the compensation subsampling scheme, and E_1 and V_1 are the corresponding operators for the initial sampling. Thus

$$V(\hat{Y}) = E_1V_2(\hat{Y}) + V_1(\hat{Y}_1). \qquad (2.3.1.2)$$

The first term is zero when $\hat{Y} = \hat{Y}_1$, a constant, as occurs when $\bar{y}_m = \bar{y}_r$; in this case $V(\hat{Y}) = V_1(\hat{Y}_1) = V(\hat{Y}_1)$. Otherwise it is positive, indicating the larger variance of compensation procedures requiring only that $E_2(\bar{y}_m) = \bar{y}_r$. We will investigate the proportionate increase in the variance of $\hat{Y}$ associated with various compensation subsampling schemes, where the proportionate increase is defined by

$$I = [V(\hat{Y}) - V(\hat{Y}_1)]/V(\hat{Y}_1) = E_1V_2(\hat{Y})/V(\hat{Y}_1).$$

The magnitude of I depends on the form of subsampling employed: it is, for instance, larger for unrestricted sampling than SRS and likely to be larger for SRS than for proportionate stratified sampling. The denominator of I is the variance of the estimator of the population total for all compensation schemes with $\bar{y}_m = \bar{y}_r$. Conditional on r, this variance is

$$V(\hat{Y}_1) = N^2V(\bar{y}_r) = N^2[1 - (r/R)]S_r^2/r \qquad (2.3.1.3)$$

where

$$S_r^2 = \Sigma(Y_{ri} - \bar{Y}_r)^2/(R - 1).$$

(a) Subsampling by SRS. Consider first the SRS weighting and imputation schemes. With either scheme, the estimator of Y can be expressed as

$$\hat{Y}_2 = N[(k + 1)r\bar{y}_r + t\bar{y}_t]/n \qquad (2.3.1.4)$$

where $\bar{y}_t$ is the mean of the SRS of t respondents chosen to have an additional weight or to be used as a donor an extra time. The variance of $\hat{Y}_2$ can be obtained from (2.3.1.2), with the second component of that equation being given by (2.3.1.3). The first component is

$$E_1V_2(\hat{Y}_2) = E_1[N^2t(1 - P)s_r^2/n^2] = N^2t(1 - P)S_r^2/n^2 \qquad (2.3.1.5)$$

where $s_r^2 = \Sigma(y_{ri} - \bar{y}_r)^2/(r - 1)$, and $P = t/r$ is the sampling fraction for the subsample of size t. Then

$$V(\hat{Y}_2) = V(\hat{Y}_1) + N^2S_r^2P(1 - P)(k + 1 + P)^{-2}/r. \qquad (2.3.1.6)$$

If the finite population correction $[1 - (r/R)]$ can be ignored, the proportionate increase in variance from using the SRS weighting or imputation scheme over the simple equal reweighting scheme (or another compensation procedure that satisfies $\bar{y}_m = \bar{y}_r$) is

$$I_2 = P(1 - P)/(k + 1 + P)^2. \qquad (2.3.1.7)$$

For fixed k the maximum increase occurs when $P = (k + 1)/(2k + 3)$ and clearly the increase is greater the smaller the value of k. When $k = 0$, i.e. the number of nonrespondents is less than the number of respondents and $t = m$, the maximum increase occurs when $P = m/r = 1/3$; in this case the proportionate increase in variance is 1/8.

As Kish (1965, Section 11.7; 1976) demonstrates, the SRS weighting scheme described above produces a more precise estimator than one with greater variability in the weights, as for instance would occur if, say, a subsample of t/2 respondents were assigned weights of (k + 3) rather than t respondents being assigned weights of (k + 2), or the equivalent with donors in an imputation scheme.

(b) Subsampling by unrestricted sampling. The SRS weighting and imputation schemes also have lower variances than if unrestricted sampling had been used. For the latter scheme, suppose that m respondents are chosen with equal probability with replacement to serve as donors for imputation or to receive unit increases in weight. The estimator of Y may then be expressed as $\hat{Y}_3 = N(y_r + y_m)/n$, with

$$E_1V_2(\hat{Y}_3) = E_1[N^2V_2(y_m)/n^2] = E_1[N^2m(r - 1)s_r^2/rn^2]$$

$$= N^2m(r - 1)S_r^2/rn^2. \quad (2.3.1.8)$$

Thus, ignoring the f.p.c. $[1 - (r/R)]$, the proportionate increase in variance over compensation schemes in which $\bar{y}_m = \bar{y}_r$ is

$$I_3 = m(r - 1)/n^2 \simeq \bar{m}(1 - \bar{m}) \quad (2.3.1.9)$$

if r is large so that $(r - 1) \simeq r$. The maximum increase in this case occurs when the nonresponse rate $\bar{m}$ is 0.5; the proportionate increase in variance is then 1/4. From (2.3.1.5) and (2.3.1.8), the increase in variance through using unrestricted sampling over SRS in a weighting or imputation scheme is

$$V(\hat{Y}_3) - V(\hat{Y}_2) = V(\hat{Y}_1)[kr(r - 1) + t(t - 1)]/n^2. \quad (2.3.1.10)$$

This difference is always nonnegative. It is zero only when $k = 0$ or $r = 1$ and $t = 0$ or 1.

(c) <u>Subsampling by proportionate stratified sampling</u>. The increase in variance (2.3.1.7) through using the SRS weighting or imputation scheme can be substantially reduced by replacing SRS by proportionate stratified sampling. For the SRS scheme, from (2.3.1.5)

$$V_2(\hat{Y}_2) = N^2t(1 - P)s_r^2/n^2.$$

The conditional variance of the estimator of the population total, $\hat{Y}_4$, with a proportionate stratified sampling scheme is

$$V_2(\hat{Y}_4) = N^2t(1 - P)s_{rw}^2/n^2$$

where s_{rw}^2 is the average within-stratum variance. The ratio of these two variances is thus

$$d = V_2(\hat{Y}_4)/V_2(\hat{Y}_2) = s_{rw}^2/s^2. \qquad (2.3.1.11)$$

If effective stratification is employed, s_{rw}^2 will be small compared with s_r^2, in which case the losses of subsampling will be substantially reduced.

The respondents can in fact be stratified highly effectively in terms of their y-values. One simple scheme is to divide them into, say, s equal-sized strata, with the first stratum containing the r/s respondents with the largest y-values, the second stratum containing the r/s respondents with the largest y-values among the remainder, etc., and then to select t/s respondents from each stratum (assuming for simplicity that r/s and t/s are integers). With this kind of stratification, which is closely related to problems of matching and grouping, the value of d rapidly becomes small as s, the number of strata, increases: for s = 5 the value of d is around 0.05 to 0.10 for many distributions of y. (See, for instance, Aigner _et al_., 1975; Anderson _et al_., 1980.) The maximum value that s can take is s = t. A systematic sample from a list of respondents ordered by their y-values should give similar gains, and this scheme is very easy to implement.

(d) _Subsampling with a larger sample._ Finally, mention should be made of another method of reducing the increases in variance arising from the SRS weighting and imputation schemes: this involves increasing the subsample size from t to, say, ct ($ct \le r$), with a counterbalancing adjustment to the additional weight to 1/c, or with splitting each nonrespondent into c parts, each with weight 1/c, and using different donors for each part. The estimator for the total for this scheme may be expressed as

$$\hat{Y}_5 = N[(k + 1)r\bar{y}_r + t\bar{y}_{ct}]/n \qquad (2.3.1.12)$$

with

$$E_1V_2(\hat{Y}_5) = E_1[N^2t(1 - cP)s_r^2/cn^2] = N^2t(1 - cP)S_r^2/cn^2.$$

In this case, the proportionate increase in variance over schemes in which $\bar{y}_m = \bar{y}_r$ is

$$I_5 = P(1 - cP)/c(k + 1 + P)^2. \qquad (2.3.1.13)$$

Comparison of this increase with that for the SRS scheme (2.3.1.7) shows that this increase is smaller by the factor (1/c) and also by a factor (1 - cP)/(1 - P). The first factor occurs because of the increased sample size and the second because of the f.p.c. term. The use of the above scheme with imputation has been proposed by Kish (1979), who has termed it a repeated replication imputation procedure (RRIP). A note he has written on the procedure is included as an appendix to this report. See also Section 5.1 on multiple imputations.

Kalton and Kish (1981) provide further discussion of the variances examined in this section. Other investigations of the effect of missing data compensation procedures on the precision of survey estimators are reported by Bailar _et al_. (1978), Ernst (1980), Ford (1980), and Scheuren (1980). The loss of precision from random duplication of records is also discussed by Roshwalb (1953) and in the texts of Hansen _et al_. (1953, Vol. II, pp. 139-141) and Kish (1965, pp. 427-428).

2.3.2 Standard Error Estimation

A second criterion for choice between alternative compensation procedures is whether they enable reasonable variance estimates to be computed in a straightforward way. At least for the simple case considered in this chapter, this criterion tends to favor weighting adjustments rather than imputation. To establish this point, note that the estimator $\hat{Y} = N\Sigma w_i y_{ri}/n$ serves to represent the estimator of the population total for all the main compensation procedures mentioned in this chapter. This is obvious for all weighting adjustments; for imputation procedures involving the assignment of respondent values to nonrespondents, the respondent values are in effect weighted by $w_i = (1 + g_i)$, where respondent i serves as donor to g_i nonrespondents; and the case where all m nonrespondents are assigned the respondent mean $\bar{y}_r$ can be viewed as giving all respondents the weight of n/r.

Consider now the estimation of the variance of $\hat{Y}$. We assume that the population is large so that the f.p.c. may be ignored and the SRS of sampled units may be treated as independently selected. We also assume that the weights w_i

associated with different respondents are uncorrelated. This assumption could be achieved by giving all respondents the same weight, by imputing the value $\bar{y}_r$ to all nonrespondents, or by allocating weights or selecting donors by unrestricted sampling. As an approximation it also covers SRS weighting and imputation schemes where the sampling fraction for the subsample, t/r, is small. With these assumptions the quantities $u_i = w_i y_{ri}$ are uncorrelated. Hence the variance of $\hat{Y} = N\Sigma u_i/n$, where the summation runs over the r respondents, is

$$V(\hat{Y}) = N^2 r\sigma_u^2/n^2 \qquad (2.3.2.1)$$

where σ_u^2 is the population variance of the U_i.

Providing the sample values u_i are available, σ_u^2 may be readily estimated by $s_u^2 = \Sigma(u_i - \bar{u})^2/(r - 1)$ to produce the variance estimator

$$v(\hat{Y}) = N^2 r s_u^2/n^2. \qquad (2.3.2.2)$$

With weighting procedures, all the respondent records automatically contain both w_i and y_{ri}, so that the u_i are obtained straightforwardly. This is, however, not the case with imputation procedures. What would be needed would be to log on the respondent's record a variable to indicate the number of times he was used as a donor (g_i) and to attach identifying flags to the nonrespondents (so that they could be eliminated for variance estimation purposes). These steps add to the complexity of the imputation procedure and, because each survey variable has to be treated separately, substantially increase the length of the data records for sampled units; since they are not required for general analyses, they are often not deemed necessary. Although imputed values are frequently flagged, this is not a universal practice; the logging of the use of respondents as donors is seldom carried out.

In passing, it is worth noting that flagging of nonrespondents and logging of the use of respondents as donors can be handled by a single variable. A simple scheme for epsem samples is to define a weight variable $w_i = (1 + g_i)$ for respondents and $w_i = 0$ for nonrespondents. The scheme can be extended to non-epsem samples by defining $w_i = a_i + \Sigma a_j$ for respondents, where a_i is the initial weight of respondent i and Σa_j is the sum of the weights of the nonrespondents for which respondent i acts as donor, and $w_i = 0$ for nonrespondents.

Often when imputation is used the researcher is confronted with a full data set of n values without being able to distinguish between actual and imputed values. A natural path to take in this situation is to treat the imputed values as actual values, but the resultant variance estimator for $\hat{Y}$, $v(\hat{Y}) = N^2s^2/n$, clearly underestimates $V(\hat{Y})$. An obvious reason for the underestimation is that the sample size is taken to be n rather than r, but there is also another factor involved. To illustrate this, we will examine the situation with two alternative imputation procedures.

First consider the imputation procedure which assigns the respondent mean to all nonrespondents. The sample estimate of the element variance from the full data set is then

$$\begin{aligned} s^2 &= \Sigma(y_i - \bar{y})^2/(n - 1) \\ &= [\Sigma(y_{ri} - \bar{y}_r) + \Sigma(y_{mi} - \bar{y}_r)^2]/(n - 1) \end{aligned}$$

where $y_{mi} = \bar{y}_r$. Hence

$$s^2 = \Sigma^r(y_{ri} - \bar{y}_r)/(n - 1) = s_r^2(r - 1)/(n - 1). \qquad (2.3.2.3)$$

Thus, since $E(s_r^2) = S_r^2$,

$$E[v(\hat{Y})] = (r - 1)N^2S_r^2/n(n - 1) = r(r - 1)V(\hat{Y})/n(n - 1). \qquad (2.3.2.4)$$

The component (r/n) is accounted for by the correction from the inflated sample size of n to the true one of r, while the second component $(r - 1)/(n - 1)$ derives from the underestimation of s^2 for S_r^2.

Secondly, consider the SRS imputation procedure. In this case s^2 is an almost unbiased estimator of S_r^2, as is shown below, but there is another factor involved in making $v(\hat{Y})$ an underestimate of $V(\hat{Y})$. To show that s^2 is almost unbiased for S_r^2, first note that

$$\Sigma^n(y_i - \bar{y})^2 = (k + 1)\Sigma^r(y_{ri} - \bar{y}_r)^2 + \Sigma^t(y_{ti} - \bar{y}_r)^2 - (t^2/n)(\bar{y}_t - \bar{y}_r)^2$$

where y_{ti} is the y-value for a respondent chosen by SRS to act as donor for an extra time and $\bar{y}_t = \Sigma y_{ti}/t$. Then

$$\begin{aligned} E_2[\Sigma^n(y_i - \bar{y})^2] &= (k + 1)(r - 1)s_r^2 + [t(r - 1)s_r^2/r] \\ &\quad - (t^2/n)[1 - (t/r)]s_r^2/t \\ &= s_r^2\{[(k + 1)r + t - 1] - k - (t/r) - (t/n)[1 - (t/r)]\} \\ &= s_r^2\{(n - 1) - k - (t/r) - (t/n)[1 - (t/r)]\}. \end{aligned}$$

Hence

$$E_1E_2(s^2) = S_r^2\{1 - (n - 1)^{-1}[k + P + (t/n)(1 - P)]\} \qquad (2.3.2.5)$$

with $P = t/r$. In the common situation when $k = 0$, i.e. $r > m$, this becomes $E_1E_2(s^2) = S_r^2[1 - 2\bar{m}/(n - 1)]$. In any case, if r is large so that terms of order $(1/r)$ may be neglected in (2.3.2.5),

$$E_1E_2(s^2) \simeq S_r^2. \qquad (2.3.2.6)$$

However, although s^2 is almost unbiased for S_r^2, the variance estimator $v(\hat{Y}) = N^2s^2/n$ underestimates $V(\hat{Y})$, which in this case is given by equation (2.3.1.6), that is

$$V(\hat{Y}) \simeq N^2S_r^2[1 + P(1 - P)(k + 1 + P)^{-2}]/r.$$

Hence

$$E[v(\hat{Y})] \simeq (r/n)[1 + P(1 - P)(k + 1 + P)^{-2}]^{-1}V(\hat{Y}). \qquad (2.3.2.7)$$

The first factor (r/n) reflects the effect of treating the sample as if it were one of size n rather than r, while the second factor reflects the increase in variance through the SRS selection of t of the r respondents to act as donors on the additional occasion.

These results show that, when the amount of nonresponse is appreciable, the treatment of imputed values for nonrespondents as if they were actual values leads to a substantial overestimation of the sample's precision. A method of obtaining suitable variance estimators when imputed values are assigned to nonrespondents, employing the idea of multiple imputation, has been proposed by Rubin (1978, 1979b, 1980). Multiple imputations are discussed in Section 5.1.

In concluding this section, it should be noted that the variance estimation procedure outlined for use with weighting adjustments applies only to the simple case in which the weighting adjustment procedure is applied across the whole sample, that sample being selected by SRS. Weighting adjustment procedures applied within subgroups of the sample are considered later in the report.

2.3.3 General Applicability

The first two criteria for choice of compensation procedure - the precision and estimable precision of the estimator - have been illustrated by the estimation of a single parameter, the population total. The aim of the general purpose compensation procedures discussed in this report is, however, to produce data sets that can be used legitimately for many forms of analysis. The third criterion is, therefore, that after compensation the data set should satisfy the requirements for general analyses. In order to examine how the various compensation procedures fare with regard to this criterion we will assume that the missing data are missing at random; the alternative assumption $\bar{Y}_r = \bar{Y}_m$ considered earlier becomes inadequate when interest extends to the estimation of parameters other than the mean and total.

After the mean and total, the two most obvious statistics to examine for univariate analyses are the distribution of the variable and its variance, the former because survey reports often contain categorized distributions and the latter because element variances play a role in many forms of analysis. The effect of compensation procedures on relationships can be most simply investigated in terms of the covariance between two variables, $S_{xy} = \Sigma(X_i - \bar{X})(Y_i - \bar{Y})/(N - 1)$. We will examine various compensation procedures in turn to see their effects on the estimation of these three statistics. For this purpose we will employ the following estimators of the element variance and covariance from a weighted sample of size n:

$$s_w^2 = [n/(n - 1)][\Sigma w_i(y_i - \bar{y})^2/\Sigma w_i]$$
$$= [n/(n - 1)][\Sigma(w_i y_i^2/\Sigma w_i) - \bar{y}^2)] \quad (2.3.3.1)$$

and

$$s_{wxy} = [n/(n - 1)][\Sigma w_i(x_i - \bar{x})(y_i - \bar{y})/\Sigma w_i]$$
$$= [n/(n - 1)][(\Sigma w_i x_i y_i/\Sigma w_i) - \bar{x}\bar{y}]. \quad (2.3.3.2)$$

These estimators are commonly used, as are also the alternative estimators in which the term $n/(n - 1)$ is dropped; the difference between these two forms of estimator is negligible for large n.

(a) Equal weighting scheme. The effects of the simple weighting procedure in which all respondents are assigned the same weight, n/r, to adjust for the nonrespondents are readily determined. The procedure clearly retains the respondents' distribution for the y-variable. The weighted sample variance estimator is

$$s_w^2 = [r/(r - 1)][\Sigma w_i(y_{ri} - \bar{y}_r)^2/\Sigma w_i]$$
$$= \Sigma(y_{ri} - \bar{y}_r)^2/(r - 1) = s_r^2 \quad (2.3.3.3)$$

which is unbiased for S_r^2. Similarly the weighted sample covariance estimator is equal to s_{rxy}, which is unbiased for S_{rxy}.

(b) SRS weighting scheme. Results for the SRS weighting procedure will be derived as a special case of any epsem subsampling scheme for assigning the weights to compensate for nonrespondents. First, consider the distribution of the y-variable in the weighted sample. Let $d_{ij} = 1$ if y_i falls in class j, i.e. $Y_{j-1} \le y_i < Y_j$, and $d_{ij} = 0$ otherwise. The weighted estimate of the proportion of the population in class j is then $p_j = \Sigma w_i d_{ij}/n$, and its conditional expectation over ways of subsampling respondents or assigning the weights is

$$E_2(p_j) = \Sigma E_2(w_i)d_{ij}/n = \Sigma(n/r)d_{ij}/n = \Sigma d_{ij}/r. \qquad (2.3.3.4)$$

Thus the conditional expectation of the weighted estimate is the unweighted respondent estimate, which is unbiased for the respondent proportion, and hence for the population proportion under the missing at random assumption.

The conditional expectation of s_w^2 for any epsem subsampling weighting procedure in which t respondents are given a weight of (k + 2) and (r - t) are given a weight of (k + 1) is

$$\begin{aligned} &E_2\{[r/(r-1)][\Sigma(w_i y_i^2/n) - \bar{y}^2]\} \\ &\qquad = [r/(r-1)][(\Sigma y_i^2/r) - E_2(\bar{y} - \bar{y}_r)^2 - \bar{y}_r^2] \\ &\qquad = s_r^2 - [r/(r-1)]t^2 V_2(\bar{y}_t)/n^2. \qquad (2.3.3.5) \end{aligned}$$

For any nonclustered epsem subsampling scheme for the t respondents to be assigned the weight of (k + 2), $V_2(\bar{y}_t)$ is of order (1/t). Hence the second term in (2.3.3.5) is of order (t/n^2). Thus, providing n is large, s_w^2 is an approximately unbiased estimator of s_r^2, which is an unbiased estimator of S_r^2 (see also Kish, 1965, Section 2.8C).

The same approach can also be applied with the weighted covariance to give

$$E_2(s_{wxy}) = s_{rxy} - [r/(r-1)]t^2 C_2(\bar{x}_t, \bar{y}_t)/n^2 \qquad (2.3.3.6)$$

where $C_2(\bar{x}_t, \bar{y}_t)$, the conditional covariance of $\bar{x}_t$ and $\bar{y}_t$, is of order (1/t). For large n, s_{wxy} is thus an approximately unbiased estimator of S_{rxy} for any epsem non-clustered subsampling scheme for assigning the weights.

Results for the SRS weighting scheme are obtained by substituting

$$V_2(\bar{y}_t) = [1 - (t/r)]s_r^2/t$$

and

$$C_2(\bar{x}_t, \bar{y}_t) = [1 - (t/r)]s_{rxy}/t$$

in (2.3.3.5) and (2.3.3.6) respectively. For the conditional expectation of s_w^2 this gives

$$E_2(s_w^2) = s_r^2\{1 - [(r - 1)/r][t(r - t)/n^2 r]\}$$

Approximating $(r - 1) \simeq r$, with $P = t/r$ being the subsampling fraction and $\bar{r} = r/n$ being the response rate,

$$E_2(s_w^2) \simeq s_r^2[1 - \{\bar{r}P(1 - P)/n\}]. \qquad (2.3.3.7)$$

A corresponding result may be obtained in the same way for the covariance.

(c) Mean-value imputation. Turning to imputation procedures, consider first the one in which the mean value for respondents $\bar{y}_r$ is assigned to all nonrespondents. This procedure distorts the y-distribution by concentrating all the nonrespondents at the one value. The proportion of the total sample in class j is

$$p_j = \Sigma d_{ij}/n = r_j/n \qquad \text{if } \bar{y}_r < Y_{j-1} \text{ or } \bar{y}_r \geq Y_j$$

and $$p_j = (r_j + m)/n \qquad \text{if } Y_{j-1} \leq \bar{y}_r < Y_j,$$

where r_j is the number of respondents falling in class j. Thus, p_j overestimates the population proportion in the class containing $\bar{y}_r$, and underestimates the proportions in all other classes. It may also be noted that if y is a discrete variable (e.g. number of children in the household), the imputed value $\bar{y}_r$ will generally be an impossible non-integer individual value (e.g. 1.7). The extreme case is when y is a 0-1 variable and $\bar{y}_r$ is the proportion of respondents with a particular attribute. By concentrating the nonrespondents at the mean, this procedure also leads to an underestimation of the element variance: as already shown in equation (2.3.2.3), $s^2 = \Sigma(y_i - \bar{y})^2/(n - 1)$ underestimates S_r^2 by a factor $(r - 1)/(n - 1)$.

This imputation procedure can be extended to cover two (or more) variables by substituting $\bar{x}_r$ and $\bar{y}_r$ for missing x- and y-values. If there are r' respondents providing both x and y values, the sample covariance from the data set incorporating the imputed values, $s_{xy} = \Sigma(x_i - \bar{x})(y_i - \bar{y})/(n - 1)$, underestimates S_{rxy} by a factor $(r' - 1)/(n - 1)$, a straightforward extension of (2.3.2.3). If both values are missing for the same m nonrespondents, the underestimation is by the factor $(r - 1)/(n - 1)$ since $r = r'$. It is interesting to note that in this latter case the correlation coefficient $r_{xy} = s_{xy}/s_x s_y$ and the regression coefficient $b_{yx} = s_{xy}/s_x^2$ from the completed data set are consistent estimators of their corresponding population parameters, because the factor $(r - 1)/(n - 1)$ cancels out between their numerators and denominators.

(d) SRS imputation scheme. Univariate results for the SRS imputation scheme, and other schemes in which the t respondents chosen to serve as donors on an additional occasion are selected by epsem subsampling, can be obtained in the same way as those for the equivalent weighting schemes. It follows immediately that the distribution of the resulting data set provides an unbiased estimator of the respondent population distribution. The complete data set element variance can be expressed as

$$\begin{aligned} s^2 &= \Sigma^n(y_i - \bar{y})^2/(n - 1) \\ &= [\Sigma^r w_i(y_{ri} - \bar{y}_r)^2 - n(\bar{y} - \bar{y}_r)^2]/(n - 1) \\ &= \{\Sigma^r w_i(y_{ri} - \bar{y}_r)^2 - [t^2(\bar{y}_t - \bar{y}_r)^2/n]\}/(n - 1) \end{aligned}$$

where w_i is the number of times respondent i serves as a donor. Hence

$$\begin{aligned} E_2(s^2) &= [n/(n - 1)][(r - 1)/r]s_r^2 \\ &\qquad - t^2[1 - (t/r)]s_r^2/tn(n - 1) \\ &= [n/(n - 1)][(r - 1)/r]E_2(s_w^2) \qquad (2.3.3.8) \end{aligned}$$

from (2.3.3.5). Thus, for large r, $E_2(s^2) \approx E_2(s_w^2)$ where s_w^2 is the estimate of the element variance from the equivalent weighting scheme. As shown earlier, s_w^2 is an approximately unbiased estimator for s_r^2 with large n, and hence s^2 is also.

The extension of the SRS or other epsem imputation scheme to pairs of variables introduces a number of possible outcomes depending on the pattern of missing data and the manner of applying the imputation scheme. One simple case is when both x and y are reported or both are missing. If each nonrespondent receives both imputed values from the same donor, the effect on the covariance is a straightforward generalization of that for the element variance (2.3.3.8), i.e.

$$E_2(s_{xy}) = [n/(n - 1)][(r - 1)/r]E_2(s_{wxy}) \tag{2.3.3.9}$$

Since, for large n, s_{wyx} is an approximately unbiased estimator of S_{rxy}, s_{xy} is also.

An alternative possibility is to assign the imputed values of x and y independently. When the x and y are missing in pairs as above, the effect of this imputation scheme on the covariance can be seen as follows:

$$\begin{aligned}\Sigma^n(x_i - \bar{x})(y_i - \bar{y}) &= \Sigma^n(x_i - \bar{x}_r)(y_i - \bar{y}_r) - n(\bar{x}_r - \bar{x})(\bar{y}_r - \bar{y})\\ &= \Sigma^r(x_{ri} - \bar{x}_r)(y_{ri} - \bar{y}_r) + \Sigma^m(x_{mi} - \bar{x}_r)(y_{mi} - \bar{y}_r)\\ &\quad - n(\bar{x}_r - \bar{x})(\bar{y}_r - \bar{y}).\end{aligned} \tag{2.3.3.10}$$

With epsem subsampling and independent assignments of the x and y values, $E_2(\bar{x}) = \bar{x}_r$, $E_2(\bar{y}) = \bar{y}_r$, $E_2(\bar{x}\bar{y}) = \bar{x}_r\bar{y}_r$ and $E_2(x_{mi} - \bar{x}_r)(y_{mi} - \bar{y}_r) = 0$. Hence

$$E_2(s_{xy}) = [(r - 1)/(n - 1)]s_{rxy}. \tag{2.3.3.11}$$

Thus, since $E(s_{rxy}) = S_{rxy}$, s_{xy} computed under this imputation procedure underestimates S_{rxy} by a factor $(r - 1)/(n - 1)$. This result is the same as that obtained for the covariance when the imputation procedure assigned the respondent mean to all nonrespondents. However, unlike that case, the factor $(r - 1)/(n - 1)$ applies here only to the covariance and not to the variances. There is therefore no cancellation of the factor between numerator and denominator of the correlation or regression coefficients: both are underestimated by the factor $(r - 1)/(n - 1)$.

A third possibility is when data on one of the variables, x, is complete, and only y has missing values. If the y_{mi}'s are assigned independently of the x_{mi} values (the reported x values for sampled elements with missing y values),

$$E_2(x_{mi} - \bar{x})(y_{mi} - \bar{y}) = 0.$$

In addition, since $\bar{x}$ does not depend on the subsampling scheme, it follows that:

$$E_2[n(\bar{x}_r - \bar{x})(\bar{y}_r - \bar{y})] = 0.$$

Hence, from (2.3.3.10),

$$E_2[\Sigma^n(x_i - \bar{x})(y_i - \bar{y})] = (r - 1)s_{rxy}.$$

Thus again s_{xy} underestimates S_{rxy} by the factor $(r - 1)/(n - 1)$.

A fourth possibility is that both x and y are subject to missing data, and that imputations are made independently for each variable. In the partition of the sum of products in (2.3.3.10), let the first summation relate to the respondents who provide values for both x and y (i.e. replace r by r'), let the second summation relate to the sample members who provide partial data or no data on x and y, let $\bar{x}_r$ denote the sample mean for <u>all</u> sample members providing x values, and let $\bar{y}_r$ denote the sample mean for all sample members providing y values. Then, with independent imputations for sample members providing no data,

$$E_2 (x_{mi} - \bar{x}_r)(y_{mi} - \bar{y}_r) = 0$$

and

$$E_2[n(\bar{x}_r - \bar{x})(\bar{y}_r - \bar{y})] = 0.$$

Thus

$$E_2[\Sigma^n(x_i - \bar{x})(y_i - \bar{y})] = \Sigma^{r'}(x_{ri} - \bar{x}_r)(y_{ri} - \bar{y}_r)$$
$$= \Sigma^{r'}(x_{ri} - \bar{x}_r')(y_{ri} - \bar{y}_r') + r'(\bar{x}_r' - \bar{x}_r)(\bar{y}_r' - \bar{y}_r)$$

where $\bar{x}_r'$ and $\bar{y}_r'$ are the means of the x and y values for the r' respondents providing both values. Hence

$$E_2(s_{xy}) = [(r' - 1)/(n - 1)]s_{rxy}' + r'(\bar{x}_r' - \bar{x}_r)(\bar{y}_r' - \bar{y}_r)/(n - 1)$$

where s_{rxy}' relates to the subset of respondents providing both x and y values. If the missing x and y values are independently missing at random, the second term is zero and s_{rxy}' estimates S_{xy}. Then s_{xy} underestimates S_{xy} by the factor $(r' - 1)/(n - 1)$.

2.4 Concluding Remarks

There is a close relationship between weighting and imputation as means of making compensations for missing data, but there are also differences between the two types of procedures. Weighting is based on matching nonrespondents to respondents in terms of the data available on the nonrespondents, and increasing the weights of the matched respondents to account for the nonrespondents. If the matching is done on an individual basis, one respondent for each nonrespondent, the weighting procedure could be equivalently carried out by assigning the matched respondent's values to the nonrespondent, i.e. by imputation. The difference between the two procedures is that, when weighting, the data set consists of r respondent records with weights adding to n, whereas with the imputation procedure the data set consists of n records, one for each respondent and nonrespondent.

Weighting need not, however, be carried out by individual matching, that is by adding integer weights to the respondents. The individual matching procedure in fact introduces an unnecessary stage of subsampling to choose the respondents for use as matched cases, and this gives rise to increased sampling error in the survey estimates. The simple weighting adjustment procedure of increasing all respondents' weights (in a defined subgroup) by the same, probably fractional, factor is a more efficient procedure because it avoids this subsampling stage. Before many computer programs had been developed to handle weighted data, a case could be made for using individual matching and the imputation scheme described above in order to avoid weights; nowadays, however, this consideration no longer applies, and the simple weighting procedure is the obvious choice when this type of adjustment is being made. The simple weighting procedure retains all the properties of the respondent sample, and hence leads to good estimates of other univariate statistics as well as means and totals and also to good estimates of multivariate statistics (under the assumption that the missing data are missing at random). It is for these reasons that the simple weighting procedure - applied separately with different weights within subgroups of the sample - is widely used to compensate for unit nonresponse.

Where there is a choice between weighting and imputation procedures, the various considerations discussed favor weighting. Thus weighting is generally employed for unit nonresponse, while imputation is reserved for item nonresponse (for which weighting is impracticable). Weighting also probably has an advantage over imputation in terms of economy of analysis of the resulting data set; it is likely to be more economical to analyze a weighted data set of r cases than an unweighted one of n cases. However, weighting adjustments require the matching of nonrespondents with respondents in terms of their available data, and this matching becomes impossible when a sizeable amount of data is available for the nonrespondents, as occurs in particular with item nonresponse. In such cases, the best that can be done is to match nonrespondents and respondents on a selection of major variables, and then to assign a matched respondent's values for the missing values in a nonrespondent's record. This imputation procedure is the same as the imputation scheme corresponding to the individual weighting scheme described above, except that not all the respondent's values are assigned to the nonrespondent, but only those for the variables for which the nonrespondent has missing values.

The use of imputation raises problems about sampling error estimation and also about the estimation of relationships between variables; these problems are discussed further in Chapter 4. The subsampling in the choice of donor adds another sampling error component, but this may be reduced by the use of multiple imputations and/or by the choice of subsampling scheme.

3. Weighting Adjustments

3.1 Types of Weighting Adjustment

The simple weighting procedure examined in the previous chapter has little practical utility since its application does not alter the estimator of the population mean ($\bar{y}_w = \bar{y}_r$) and the population total can be more easily estimated by $N\bar{y}_r$ than by Fy_w ($N\bar{y}_r = Fy_w$). The procedure, however, forms the basis of more general weighting procedures, in which it is applied separately to different subgroups of the sample. These subgroups will henceforth be called weighting classes or simply classes; some researchers call them adjustment cells.

In Chapter 2 the weighting factor applied with the simple weighting adjustment procedure was the inverse of the response rate. This scheme can be extended to weighting classes by applying the inverse of the response rate in a class as the weighting factor for that class. The effect of this kind of adjustment is to make the distribution of the weighted sample of respondents across the weighting classes the same as that for the total sample of respondents and nonrespondents. For this reason, the weights will be called sample weights. Sample weighting adjustments are discussed in Section 3.3.

Sometimes the population totals for the weighting classes are known, in which case the weighting factors can be based on them rather than on the sample data. These weights will be called population weights, and population weighting adjustments are discussed in Section 3.2.

When the weighting classes are formed from several characteristics, the number of classes may be large. The application of weighting adjustments independently from one class to another can then result in substantial variability in the weights used and hence in sizeable increases in variances for the survey estimates. The technique of raking ratio estimation discussed in Section 3.4 can be useful in this situation.

It may sometimes be reasonable to assume that nonrespondents have similar characteristics to those respondents whose cooperation in the survey was difficult to secure. If so, the nonresponse weighting adjustments may be restricted to this kind of respondent. Weighting procedures of this type are discussed in Section 3.5.

The final section of the chapter discusses some general considerations in weighting adjustments.

3.2 Population Weighting Adjustments

The application of population weighting adjustments requires that information be available on the distribution of the survey population over the weighting classes. This information may, for instance, come from the previous Census. The data from the respondent sample is then weighted so that its weighted distribution over the classes conforms to that from the external source. As an example, an adjustment is employed with the Current Population Survey (CPS) to bring the survey into line with independent estimates of the U.S. population by age, sex and race, these estimates being obtained by updating the previous Census taking account of births, deaths, aging and migration (Bailar *et al.*, 1977). It should be recognized that the data from the external source will themselves be imperfect in some degree. To the extent that this is so the weighted sample data will not truly represent the distribution of the survey population over the weighting classes. Population weighting adjustments bear a close resemblance to post-stratification (stratification after selection), and are sometimes called post-stratification adjustments. Since there are some differences between the two procedures we prefer to employ the term population weighting adjustments when the respondent sample is weighted to conform to data from an external source.

Two differences between population and sample weighting adjustments deserve mention. First, there is a difference in the data requirements. As noted, population weighting adjustments require knowledge of the population distribution over the weighting classes, but sample weighting adjustments do not. On the other hand, sample weighting adjustments require that both respondents and nonrespondents are divided into weighting classes, whereas population weighting adjustments require only that respondents are so divided. In other words, the characteristics defining the weighting classes need to be available for respondents and nonrespondents for sample weighting adjustments, but they need be available only for the respondents for population weighting adjustments. The second difference between the two forms of weighting adjustment relates to the type of nonresponse for which they compensate: population weighting adjustments compensate for both noncoverage and unit nonresponse, while sample weighting adjustments compensate only for unit nonresponse. Even when sample weighting adjustments are being used, a good case can be made for employing population weighting adjustments as well in order to deal with noncoverage.

To illustrate the effect of population weighting adjustments on non-response bias, we will consider - following Thomsen (1973) - the estimation of a population mean from an epsem sample in which different classes have differing response rates. Let the population be divided into H classes with N_h units in class h, let $N_h = R_h + M_h$, where R_h and M_h are the numbers of respondents and nonrespondents respectively in class h (with the assumption that respondents always respond and nonrespondents never do), and let $\bar{R}_h = R_h/N_h$ and $\bar{M}_h = M_h/N_h$ be the expected response and nonresponse rates for class h. The population mean for respondents is $\bar{Y}_r = \Sigma Y_{ri}/R = \Sigma\Sigma Y_{rhi}/\Sigma R_h$ and that for nonrespondents is $\bar{Y}_m = \Sigma Y_{mi}/M = \Sigma\Sigma Y_{mhi}/\Sigma M_h$. The overall population mean, the quantity to be estimated, is

$$\begin{aligned}\bar{Y} &= \Sigma N_h\bar{Y}_h/N = \Sigma W_h\bar{Y}_h = \Sigma W_h(\bar{R}_h\bar{Y}_{rh} + \bar{M}_h\bar{Y}_{mh}) \\ &= \Sigma W_h\bar{Y}_{rh} - \Sigma W_h\bar{M}_h(\bar{Y}_{rh} - \bar{Y}_{mh}).\end{aligned} \quad (3.2.1)$$

If the unweighted sample mean $\bar{y}_r = \Sigma\Sigma y_{rhi}/r$ is used to estimate $\bar{Y}$ its bias is

$$\begin{aligned}B(\bar{y}_r) &= \bar{Y}_r - \bar{Y} \\ &= [\Sigma\bar{R}_h W_h\bar{Y}_{rh}/\bar{R}] - \Sigma W_h\bar{Y}_{rh} + \Sigma W_h\bar{M}_h(\bar{Y}_{rh} - \bar{Y}_{mh}) \\ &= [\Sigma W_h\bar{Y}_{rh}(\bar{R}_h - \bar{R})/\bar{R}] + \Sigma W_h\bar{M}_h(\bar{Y}_{rh} - \bar{Y}_{mh}).\end{aligned} \quad (3.2.2)$$

Since $\Sigma W_h(\bar{R}_h - \bar{R}) = 0$, this bias may also be expressed as

$$B(\bar{y}_r) = [\Sigma W_h(\bar{Y}_{rh} - \bar{Y}_r)(\bar{R}_h - \bar{R})/\bar{R}] + \Sigma W_h \bar{M}_h(\bar{Y}_{rh} - \bar{Y}_{mh}) \quad (3.2.3)$$

$$= A + B. \quad (3.2.4)$$

The estimator of $\bar{Y}$ using the population weighting adjustment is $\bar{y}_p = \Sigma W_h \bar{y}_{rh}$. This estimator may be expressed as a weighted estimator $\bar{y}_p = \Sigma\Sigma w_{hi} y_{hi}/\Sigma\Sigma w_{hi}$, where the weights $w_{hi} = w_h \propto W_h/r_h$ compensate for the variation in response rates across the classes. Conditional on $r_h > 0$ for all h, the bias of $\bar{y}_p$ is given by $B(\bar{y}_p) = \Sigma W_h \bar{Y}_{rh} - \bar{Y}$. Using (3.2.1) it can be seen that

$$B(\bar{y}_p) = \Sigma W_h \bar{M}_h(\bar{Y}_{rh} - \bar{Y}_{mh}) = B. \quad (3.2.5)$$

The unweighted mean $\bar{y}_r$ is unbiased for $\bar{Y}$ if $\bar{Y}_r = \bar{Y}_m$, while the weighted mean $\bar{y}_p$ is unbiased if $\bar{Y}_{rh} = \bar{Y}_{mh}$. Equation (3.2.2) partitions the bias of $\bar{y}_r$ into two components, one arising from the variation in response rates across classes (A) and the other arising from the differences between respondent and nonrespondent means within classes (B). The use of $\bar{y}_p$ removes the first component of bias, leaving only B. If the components A and B are of the same sign, $\bar{y}_p$ cannot have greater absolute bias than $\bar{y}_r$. However, if A and B are of different signs, it is possible for $\bar{y}_p$ to have a greater absolute bias than $\bar{y}_r$; when A and B have different signs, the population weighting adjustment reduces the bias only when $|A| > 2|B|$.

The maximum amount of bias reduction through using $\bar{y}_p$ rather than $\bar{y}_r$ is $|A|$. As expressed in (3.2.3), A can be seen to be a covariance-type term between class means and response rates. The reason for expressing A in this form is to bring out the two conditions needed for it to be non-zero: (1) that the response rates should differ between classes and (2) that the respondent means should differ between classes.

The resemblance of $\bar{y}_p$ to a post-stratified mean $\bar{y}_{ps}$ is a useful connection to make, and it will be employed below in deriving the variance of $\bar{y}_p$. There are, however, two important differences between the two estimators that should be noted. In the first place, with post-stratification the strata means $\bar{y}_h$ are unbiased estimators of $\bar{Y}_h$ whereas, with population weighting adjustments, the class sample means $\bar{y}_h$ are unbiased estimators of $\bar{Y}_{rh}$, the respondent class means; the class sample means are only unbiased estimators of the overall class means if the expected values of the respondent and nonrespondent class means are equal. Secondly, the post-stratification weighting adjustment serves to correct for random sampling variability in the spread of the sample over the strata, whereas the population weighting adjustment corrects both for this random sampling variability and for the variation in response rates, and is primarily concerned with the latter. In most cases, post-stratification involves relatively minor weighting adjustments and may be thought of as a fine tuning. The weighting adjustments in $\bar{y}_p$ will often be much more variable, and hence have a greater impact on the estimator.

This second difference has consequences for the variances of the two estimators. With a SRS of size n, the variance of the post-stratified mean is approximately

$$V(\bar{y}_{ps}) \simeq (1 - f)\{\Sigma W_h S_h^2[1 + (1 - W_h)/E(n_h)]\}/n \qquad (3.2.6)$$

(Kish, 1965, p. 90). If the expected sample sizes from the strata $E(n_h) = nW_h$, are large - say greater than 10 or 20 - the term in $(1 - W_h)/E(n_h)$ is negligible so that $V(\bar{y}_{ps})$ is approximately the same as that of the mean based on a proportionate stratified sample, and with large strata this latter variance cannot exceed that of the mean from a SRS of the same size. Under similar conditions, $V(\bar{y}_p)$ may, however, exceed $V(\bar{y}_r)$, as is shown below.

Since, conditional on r_h, $E_2(\bar{y}_p) = \Sigma W_h \bar{Y}_{rh}$, a constant, $V_1 E_2(\bar{y}_p) = 0$ so that

$$V(\bar{y}_p) = E_1 V_2(\bar{y}_p) = E_1[\Sigma W_h^2(1 - f_h)S_{rh}^2/r_h]. \qquad (3.2.7)$$

Ignoring the f.p.c. term, conditional on r, and using the approximation (Stephan, 1945)

$$E_1(1/r_h) \simeq (1/rW_{rh}) + (1 - W_{rh})/r^2W_{rh}^2, \qquad (3.2.8)$$

$$V(\bar{y}_p) \simeq (\Sigma W_h^2 S_{rh}^2/W_{rh})/r + \Sigma(1 - W_{rh})(W_h/W_{rh})^2 S_{rh}^2/r^2. \qquad (3.2.9)$$

Substituting $W_h/W_{rh} = \bar{R}/\bar{R}_h$,

$$V(\bar{y}_p) \simeq \Sigma W_{rh}(\bar{R}/\bar{R}_h)^2 S_{rh}^2/r + \Sigma(1 - W_{rh})(\bar{R}/\bar{R}_h)^2 S_{rh}^2/r^2. \qquad (3.2.10)$$

This variance can then be expressed as

$$V(\bar{y}_p) \simeq \Sigma W_{rh}(\bar{R}/\bar{R}_h)^2 S_{rh}^2[1 + (1 - W_{rh})/E_1(r_h)]/r \qquad (3.2.11)$$

on substituting $r \simeq E_1(r_h)/W_{rh}$ in the second term of (3.2.10). The form of $V(\bar{y}_p)$ in (3.2.11) corresponds to that of $V(\bar{y}_{ps})$ in (3.2.6). In the same way, then, the second term in the parentheses may be neglected if the expected numbers of respondents, $E_1(r_h)$, are large for all classes, to give the approximate result:

$$V(\bar{y}_p) \simeq \Sigma W_{rh}(\bar{R}/\bar{R}_h)^2 S_{rh}^2/r \qquad (3.2.12)$$

or equivalently

$$V(\bar{y}_p) \simeq \Sigma W_h(\bar{R}/\bar{R}_h) S_{rh}^2/r. \qquad (3.2.13)$$

In order to compare $V(\bar{y}_p)$ with $V(\bar{y}_r)$, the latter may be expressed, without the f.p.c., as

$$V(\bar{y}_r) \simeq S_r^2/r \simeq [\Sigma W_{hr}(\bar{Y}_{hr} - \bar{Y}_r)^2 + \Sigma W_{rh} S_{rh}^2]/r. \qquad (3.2.14)$$

Hence

$$V(\bar{y}_r) - V(\bar{y}_p) \simeq \Sigma W_{hr}(\bar{Y}_{hr} - \bar{Y}_r)^2/r - \Sigma W_{rh} S_{rh}^2[(\bar{R}/\bar{R}_h)^2 - 1]/r \qquad (3.2.15)$$

$$= C - D.$$

No general results on the sign of the difference C - D can be established, but some special cases are of interest. First, suppose that the response rate is constant across classes, $\bar{R}_h = \bar{R}$. Then D = 0, and since C is nonnegative the difference is nonnegative, that is $\bar{y}_p$ is more precise than $\bar{y}_r$: this case corresponds to the post-stratified sampling result. Secondly, suppose that the element variances within the classes are equal, say $S_{rh}^2 = S_{r.}^2$. Then

$$D = S_{r.}^2 [\Sigma W_{rh}(\bar{R}/\bar{R}_h)^2 - 1]/r = S_{r.}^2 [(\Sigma W_h^2/W_{rh}) - 1]/r.$$

Now let $W_h = W_{rh} + a_h$, with $\Sigma W_h = \Sigma W_{rh} = 1$ and $\Sigma a_h = 0$. Then

$$\Sigma W_h^2/W_{rh} = \Sigma W_{rh} + 2\Sigma a_h + \Sigma a_h^2/W_{rh}$$

$$= 1 + \Sigma a_h^2/W_{rh} \geq 1.$$

Thus $D \geq 0$. In this case, both C and D are nonnegative. It is entirely possible for D to exceed C, as when classes differ substantially in their response rates but only slightly in their means; when this occurs, the weighted mean has larger variance than the unweighted one. Thirdly, suppose that the class means and element variances are equal, $\bar{Y}_{hr} = \bar{Y}_r$ and $S_{rh}^2 = S_{r.}^2$. Then $C = 0$ and, if the response rates differ, the difference is negative, that is $\bar{y}_p$ is less precise than $\bar{y}_r$. This result shows that the population weighting adjustment is harmful to precision when the class means are equal.

Since both $\bar{y}_p$ and $\bar{y}_r$ are biased estimators of $\bar{Y}$, it is more appropriate to compare their mean square errors (MSE's) than their variances. From earlier results, it then follows that

$$\text{MSE}(\bar{y}_r) - \text{MSE}(\bar{y}_p) = C - D + (A + B)^2 - B^2.$$

The sign and magnitude of this difference depend on the several parameters involved, so again no general conclusions can be reached. Note, however, that when the class means are equal, $\bar{Y}_{hr} = \bar{Y}_r$, $A = C = 0$, so that

$$\text{MSE}(\bar{y}_p) - \text{MSE}(\bar{y}_r) = D.$$

If the class element variances are also equal, $D \geq 0$, as shown above, so the population weighted estimator has larger MSE than the unweighted estimator in this case.

The precision of $\bar{y}_p$ obtained from a particular sample may be measured by its variance conditional on the r_h for that sample, i.e. by $V_2(\bar{y}_p)$ as given in parentheses in (3.2.7). An unbiased estimator for $V_2(\bar{y}_p)$ may then be calculated from

$$v(\bar{y}_p) = \Sigma W_h^2(1 - f_h)s_{rh}^2/r_h,$$

where $s_{rh}^2 = \Sigma(y_{rhi} - \bar{y}_{rh})^2/(r_h - 1)$. Since $V(\bar{y}_p) = E_1V_2(\bar{y}_p)$, it then follows that $v(\bar{y}_p)$ is also unbiased for $V(\bar{y}_p)$.

A variance estimator for a weighted sample was obtained in Section 2.3.2 by defining quantities $u_i = w_iy_{ri}$, and treating the u_i as independent. In the present case, the u_i are not independent, but the approach may nevertheless still yield a reasonable variance estimator; if so, it would provide a simpler variance estimator than $v(\bar{y}_p)$ above. To investigate this possibility, let the sample mean be $\bar{y}_p = \Sigma\Sigma w_{hi}y_{rhi}/r = \Sigma\Sigma u_{hi}/r = \bar{u}$, where the weights $w_{hi} = rW_h/r_h = w_h$ sum to r, with an average weight of 1. Treating the u_{hi} as independent, $V(\bar{y}_p) = \sigma_u^2/r$ may be estimated by s_u^2/r where $s_u^2 = \Sigma\Sigma(u_{hi} - \bar{u})^2/(r - 1)$. Now

$$\begin{aligned}\Sigma\Sigma(u_{hi} - \bar{u})^2 &= \Sigma\Sigma(u_{hi} - \bar{u}_h)^2 + \Sigma r_h(\bar{u}_h - \bar{u})^2 \\ &= r^2\Sigma\Sigma W_h^2(y_{hi} - \bar{y}_h)^2/r_h^2 + \Sigma r_h(w_h\bar{y}_h - \bar{y}_p)^2.\end{aligned}$$

Approximating $r \simeq (r - 1)$, $r_h \simeq (r_h - 1)$, and ignoring the f.p.c. term,

$$\begin{aligned}s_u^2/r &\simeq v(\bar{y}_p) + \Sigma r_h(w_h\bar{y}_h - \bar{y}_p)^2/r^2 \\ &= v(\bar{y}_p) + (G/r) \qquad (3.2.16)\end{aligned}$$

where $G = \Sigma r_h(w_h\bar{y}_h - \bar{y}_p)^2/r$ is the average squared deviation of the quantities $w_h\bar{y}_h$ around their mean. If the quantities $w_h\bar{y}_h$ vary considerably, s_u^2/r may substantially overestimate $V(\bar{y}_p)$. The use of s_u^2/r cannot therefore be generally recommended.

3.3 Sample Weighting Adjustments

With sample weighting adjustments, the total sample of both respondents and nonrespondents is divided into weighting classes, and the sample weight for a class is made proportional to the inverse of the class response rate. The choice of characteristics for defining the weighting classes is limited by the requirement that information on these characteristics must be available for nonrespondents as well as for respondents. One set of characteristics routinely available for both respondents and nonrespondents comprises those used in the sample design, such as the various sampling stage units and strata; weighting classes are often defined in terms of combinations of these characteristics. Thus, for instance, weighting classes can readily be formed in terms of geographical region, degree of urbanicity, etc. Platek et al. (1978) term weighting adjustments based on sample design characteristics "design-dependent balancing factors". In addition, other information may be available about the nonrespondents, perhaps from the sampling frame or perhaps collected by interviewers' observations; with careful training, interviewers can for instance collect almost complete information on the race of a nonresponding household.

With population weighting adjustments, the population mean is estimated by $\bar{y}_p = \Sigma W_h \bar{y}_{rh}$. With sample weighting adjustments the W_h are unknown, so they are estimated from the total sample as $w_h = n_h/n$. Thus the estimator of the population mean is $\bar{y}_s = \Sigma w_h \bar{y}_{rh}$, which may be alternatively expressed as $\bar{y}_s = \Sigma\Sigma w_{hi} y_{rhi} / \Sigma\Sigma w_{hi}$ where $w_{hi} \propto w_h/r_h$ rather than $w_{hi} \propto W_h/r_h$ as with population weighting adjustments.

The expected value of $\bar{y}_s$ is given by

$$E(\bar{y}_s) = E_1 E_2(\bar{y}_s) = E_1(\Sigma w_h \bar{Y}_{rh}) = \Sigma W_h \bar{Y}_{rh}$$

where E_2 is the expectation conditional on the r_h (assuming $r_h > 0$ for all classes). Since $E(\bar{y}_s) = E(\bar{y}_p)$, the bias of $\bar{y}_s$ is also given by (3.2.5), i.e. $B(\bar{y}_s) = B$, and the comparison made between the biases of $\bar{y}_p$ and $\bar{y}_r$ also applies for the biases of $\bar{y}_s$ and $\bar{y}_r$.

As population weighting adjustments may be likened to post-stratification, so sample weighting adjustments may be likened to two-phase (or double sampling) stratified estimation. The total sample of size n can be viewed as the first-phase sample, and the r respondents as the second-phase sample. The two-phase stratified mean corresponding to $\bar{y}_s$ is also of the form $\Sigma w_h \bar{y}_{rh}$, where $w_h = n_h/n$ is an estimate of W_h based on a large initial sample of size n, and where $\bar{y}_{rh} = \Sigma y_{hi}/r_h$ is the mean of the second phase SRS of size r_h selected from the n_h units in stratum h in the first-phase sample. One difference between sample weighting adjustments and two-phase stratified estimation is that the r_h units are drawn at random in stratum h with the two-phase sample design, but they are the self-selected group of respondents within the weighting class with the sample weighting adjustment procedure. In consequence, with the two-phase design, $\bar{y}_{rh}$ is unbiased for $\bar{Y}_h$, but within the weighting class $\bar{y}_{rh}$ is unbiased only for $\bar{Y}_{rh}$. A second difference is that the sample designer is able to specify the subsampling rate r_h/n_h with the two-phase design, but ordinarily he is unable to choose the response rate within a weighting class.

The correspondence between sample weighting adjustments and two-phase stratified estimation may be used in deriving the variance of $\bar{y}_s$. This derivation, unlike earlier ones, will not be made conditional on r; since $r = \Sigma r_h$ depends on the n_h which are subject to sampling variability and which appear in the estimator, it is inappropriate to take r as fixed. A '*' will be used in $V^*(\bar{y}_s)$ to indicate the unconditional variance. Three levels of expectation are used $E = E_1E_2E_3$: E_3 is expectation conditional on r_h and n_h, E_2 is expectation conditional on n_h, and E_1 is expectation over n_h. Then

$$V^*(\bar{y}_s) = E_1E_2V_3(\bar{y}_s) + E_1V_2E_3(\bar{y}_s) + V_1E_2E_3(\bar{y}_s). \qquad (3.3.1)$$

The three terms in (3.3.1) may be evaluated as follows. First, by treating the r_h respondents in class h as a SRS from the R_h population units in that class, it follows that

$$V_3(\bar{y}_s) = \Sigma w_h^2 S_{rh}^2 / r_h,$$

ignoring the f.p.c. term $[1 - (r_h/R_h)]$. Now, approximately from (3.2.8),

$$E_2(1/r_h) \simeq (1/n_h\bar{R}_h) + (1 - \bar{R}_h)/n_h^2\bar{R}_h^2 \qquad (3.3.2)$$

in which $w_h n$ may be substituted for n_h. Thus

$$E_2V_3(\bar{y}_s) \simeq (\Sigma w_h S_{rh}^2/\bar{R}_h n) + [\Sigma(1-\bar{R}_h)S_{rh}^2/\bar{R}_h^2 n^2]$$

and

$$E_1E_2V_3(\bar{y}_s) \simeq (\Sigma W_h S_{rh}^2/\bar{R}_h n) + [\Sigma(1 - \bar{R}_h)S_{rh}^2/\bar{R}_h^2 n^2]. \qquad (3.3.3)$$

Next consider the second term in (3.3.1). Since $E_3(\bar{y}_s) = \Sigma w_h \bar{Y}_{rh}$, a constant when the n_h are fixed, $V_2E_3(\bar{y}_s) = 0$, so that

$$E_1V_2E_3(\bar{y}_s) = 0. \qquad (3.3.4)$$

Finally, $E_2E_3(\bar{y}_s) = \Sigma w_h \bar{Y}_{rh}$ so that

$$V_1E_2E_3(\bar{y}_s) = \Sigma\bar{Y}_{rh}^2 V(w_h) + 2\sum_{i<j}\sum \bar{Y}_{rh}\bar{Y}_{rj}C(w_h, w_j),$$

where $C(w_h, w_j) \simeq - W_hW_j/n$ is the covariance of w_h and w_j for the initial SRS of size n, ignoring the f.p.c. $[1 - (n/N)]$. Thus

$$\begin{aligned} V_1E_2E_3(\bar{y}_s) &\simeq [\Sigma\bar{Y}_{rh}^2 W_h(1 - W_h) - 2\sum_{i<j}\sum \bar{Y}_{rh}\bar{Y}_{rj}W_hW_j]/n \\ &= \Sigma W_h(\bar{Y}_{rh} - \bar{Y}_s)^2/n. \end{aligned} \qquad (3.3.5)$$

where $\bar{Y}_s = \Sigma W_h \bar{Y}_{rh}$.

Combining (3.3.3), (3.3.4) and (3.3.5) gives

$$\begin{aligned} V^*(\bar{y}_s) \simeq\ &(\Sigma W_h S_{rh}^2/\bar{R}_h n) + [\Sigma(1 - \bar{R}_h)S_{rh}^2/\bar{R}_h^2 n^2] \\ &+ \Sigma W_h(\bar{Y}_{rh} - \bar{Y}_s)^2/n. \end{aligned} \qquad (3.3.6)$$

As discussed earlier in the case of post-stratification, the second term may be neglected if the r_h are large in every class; neglecting this term is equivalent to dropping the second term in (3.3.2), i.e. approximating $E_2(r_h^{-1})$ by $(n_h\bar{R}_h)^{-1}$.

To compare $V^*(\bar{y}_s)$ with $V^*(\bar{y}_p)$, the unconditional variance of $\bar{y}_p$ is needed. This unconditional variance may be obtained with two levels of expectation $E^* = E_1E_2$, where E_2 is conditional on r_h and E_1 is expectation over r_h without the restriction that $\Sigma r_h = r$ is a constant. Since $E_2(\bar{y}_p) = \bar{Y}_s$ is a constant, $V_1E_2(\bar{y}_p) = 0$. Thus $V^*(\bar{y}_p) = E_1V_2(\bar{y}_p)$, where $V_2(\bar{y}_p) \simeq \Sigma W_h^2 S_{rh}^2/r_h$, ignoring the f.p.c. term. Then using the approximation $E_1(1/r_h) \simeq nW_h\bar{R}_h$ for large r_h,

$$V^*(\bar{y}_p) \simeq \Sigma W_h S_h^2/\bar{R}_h n$$

which is the first term in (3.3.6). To the order of approximation used, the second term in (3.3.6) is negligible, so that $V^*(\bar{y}_s)$ exceeds the unconditional variance of $\bar{y}_p$ by the (nonnegative) third term. This result corresponds to one obtained for two-phase sampling in texts such as Raj (1968) and Cochran (1977). Since $\bar{y}_p$ and $\bar{y}_s$ have the same bias in estimating $\bar{Y}$, but $V^*(\bar{y}_s) \geq V^*(\bar{y}_p)$, it follows that $MSE^*(\bar{y}_s) \geq MSE^*(\bar{y}_p)$. There is therefore a greater likelihood that $MSE^*(\bar{y}_s)$ is larger than $MSE^*(\bar{y}_r)$ than is the case with $MSE^*(\bar{y}_p)$ (see the previous section).

To compare $V^*(\bar{y}_s)$ with $V^*(\bar{y}_r)$, the unconditional variance of $\bar{y}_r$ is needed. To derive $V^*(\bar{y}_r)$ let $E^* = E_1E_2$, where E_2 is conditional on r and E_1 is expectation over r. Since $E_2(\bar{y}_r) = \bar{Y}_r$ is a constant, $V_1E_2(\bar{y}_r) = 0$. Thus $V^*(\bar{y}_r) = E_1V_2(\bar{y}_r) \simeq E_1(S_r^2/r)$, ignoring the f.p.c. term. Using the approximation $E_1(1/r) \simeq 1/n\bar{R}$ and writing

$$S_r^2 \simeq \Sigma W_{rh}(\bar{Y}_{rh} - \bar{Y}_r)^2 + \Sigma W_{rh}S_{rh}^2,$$

$V^*(\bar{y}_r)$ may be expressed as

$$V^*(\bar{y}_r) \simeq [\Sigma W_{rh}(\bar{Y}_{rh} - \bar{Y}_r)^2 + \Sigma W_{rh}S_{rh}^2]/n\bar{R}. \qquad (3.3.7)$$

As was the case with the comparison of $V(\bar{y}_p)$ and $V(\bar{y}_r)$, no general conclusions on the sign or magnitude of $V^*(\bar{y}_s) - V^*(\bar{y}_r)$ can be reached.

In the special case that $\bar{Y}_{rh}$ and S^2_{rh} are constant for all classes, $\bar{Y}_{rh} = \bar{Y}_r = \bar{Y}_s$ and $S^2_{rh} = S^2_{r.}$, and the sample is large so that the second term in $V^*(\bar{y}_s)$ in (3.3.6) can be ignored,

$$V^*(\bar{y}_s) - V^*(\bar{y}_r) \simeq S^2_{r.}\Sigma[(W_h/\bar{R}_h) - W_{rh}/\bar{R}]/n = S^2_{r.}T/n.$$

This difference may be shown to be nonnegative by considering the summation term T, which may be expressed as

$$T = \Sigma[(N_h^2/R_hN) - (R_hN/R^2)] = (\Sigma N_h^2/R_h)/N - (N/R).$$

Since, by the Cauchy-Schwarz inequality,

$$(\Sigma R_h)(\Sigma N_h^2/R_h) \geq (\Sigma N_h)^2$$

i.e.

$$(\Sigma N_h^2/R_h) \geq N^2/R,$$

it follows that T is nonnegative. Thus, not surprisingly under these particular conditions, the use of variable weights in $\bar{y}_s$ leads to the result that $V^*(\bar{y}_s) \geq V^*(\bar{y}_r)$.

In another special case, where the response rate is constant across classes, $\bar{R}_h = \bar{R}$, $W_{rh} = W_h$ and $\bar{Y}_r = \bar{Y}_s$, so that the difference in variances of $\bar{y}_r$ and $\bar{y}_s$ reduces to

$$V^*(\bar{y}_r) - V^*(\bar{y}_s) = [(n\bar{R})^{-1} - n^{-1}]\Sigma W_{rh}(\bar{Y}_{rh} - \bar{Y}_r)^2$$
$$= [E_1(r^{-1}) - n^{-1}]\Sigma W_{rh}(\bar{Y}_{rh} - \bar{Y}_r)^2$$

which corresponds to the gains of two-phase sampling with proportionate stratification at the second phase.

The variance of $\bar{y}_s$ may be estimated by the two-phase stratified sample variance estimator. A good approximate estimator for large samples is

$$v(\bar{y}_s) = (\Sigma w_h^2 s_{rh}^2/r_h) + \Sigma w_h(\bar{y}_h - \bar{y}_s)^2/n \qquad (3.3.8)$$

(see, for instance, Kish, 1965, equation 12.1.4). It is readily apparent that

$$E_3(\Sigma w_h^2 s_{rh}^2/r_h) = \Sigma w_h^2 S_{rh}^2/r_h = V_3(\bar{y}_s)$$

so that this term is an unbiased estimator of $E_1E_2V_3(\bar{y}_s)$. The second term in (3.3.8) is a reasonable estimator of the other component of $V(\bar{y}_s)$, that is of

$$V_1E_2E_3(\bar{y}_s) = \Sigma W_h(\bar{Y}_{rh} - \bar{Y}_s)^2/n.$$

This may be demonstrated as follows:

$$\begin{aligned} E_1E_2E_3[\Sigma w_h(\bar{y}_{rh} - \bar{y}_s)^2] &= E_1E_2E_3[\Sigma w_h\bar{y}_{rh}^2 - \bar{y}_s^2] \\ &= E_1E_2[\Sigma w_h(V_3(\bar{y}_{rh}) + \bar{Y}_{rh}^2)] - V^*(\bar{y}_s) - \bar{Y}_s^2 \\ &= \Sigma W_h(\bar{Y}_{rh} - \bar{Y}_s)^2 - V^*(\bar{y}_s) + E_1E_2[\Sigma w_h S_{rh}^2/r_h] \\ &= \Sigma W_h(\bar{Y}_{rh} - \bar{Y}_s)^2 - V^*(\bar{y}_s) + \Sigma(S_{rh}^2/\bar{R}_h)/n \end{aligned}$$

approximating $E_2(1/r_h) \simeq w_h n\bar{R}_h$. For large n, the last two terms are negligible compared with the first, thus demonstrating that $\Sigma w_h(\bar{y}_{rh} - \bar{y}_s)^2$ provides an approximate estimator of $\Sigma W_h(\bar{Y}_{rh} - \bar{Y}_s)^2$. Hence $v(\bar{y}_s)$ in (3.3.8) serves as a reasonable estimator of $V^*(\bar{y}_s)$.

3.4 Raking Ratio Estimation

Raking ratio estimation, or raking, can be a useful weighting adjustment procedure when the weighting classes are defined in terms of the crosstabulation of a number of characteristics. The underlying basis of the technique is to make the weighted marginal distributions of the weighting class characteristics in the sample conform to those in the population, without ensuring that the weighted sample and population joint distributions conform. Raking was first introduced by Deming and Stephan (1940) - see also Deming (1943). It has subsequently been widely used under the name of iterative proportional fitting (IPF) in contingency table analysis, and recently has been applied by Purcell and Kish (1980) in small domain estimation.

There are two main reasons for using raking rather than separate class weighting adjustments. First, when many classes are formed by the crosstabulations of characteristics, the resulting separate class weighting adjustments can be extremely variable and unstable because of the sparse spread of the sample over the classes; raking avoids this problem. Secondly, separate class weighting adjustments require knowledge of the joint distribution of the weighting class characteristics for the population, whereas raking only requires knowledge of the population marginal distributions. Many cases occur where population marginal distributions are known, but the joint distribution is unknown, so that raking is possible while separate weighting adjustments are not.

Raking can be applied with many characteristics but for ease of exposition we will consider just two, one with H and the other with K categories. The population and sample joint distributions for these two characteristics are set out in the two bivariate tables below. The population joint distribution over the two characteristics is given by W_{hk}, with marginal distributions $W_{h.}$ and $W_{.k}$, and the corresponding sample distributions are q_{hk}, and $q_{h.}$ and $q_{.k}$. With an epsem sample $q_{hk} = n_{hk}/n$. The objective of raking is to introduce weights to make the weighted sample marginal distributions equal to $W_{h.}$ and $W_{.k}$.

Population

	1	2	...	K	Total
1	W_{11}	W_{12}	...	W_{1K}	$W_{1.}$
2	W_{21}	W_{22}	...	W_{2K}	$W_{2.}$
⋮	⋮	⋮		⋮	⋮
H	W_{H1}	W_{H2}	...	W_{HK}	$W_{H.}$
Total	$W_{.1}$	$W_{.2}$	...	$W_{.K}$	1

Sample

	1	2	...	K	Total
1	q_{11}	q_{12}	...	q_{1K}	$q_{1.}$
2	q_{21}	q_{22}	...	q_{2K}	$q_{2.}$
⋮	⋮	⋮		⋮	⋮
H	q_{H1}	q_{H2}	...	q_{HK}	$q_{H.}$
Total	$q_{.1}$	$q_{.2}$	...	$q_{.K}$	1

Raking employs an iterative procedure to meet this objective; successive modifications are made to the weights until the procedure stabilizes. One algorithm for the procedure is as follows. First, weight each cell in row h (h = 1,2,...H) by the factor $W_{h.}/q_{h.}$, so that the weighted column marginal distribution becomes $W_{h.}$; as a result of these weights, the sample marginal row distribution has become $\Sigma_h q_{hk} W_{h.}/q_{h.} = \Sigma q'_{hk} = q'_{.k}$. Next, weight the sampled units in each cell in column k (k = 1,2,...,K) by the factor $W_{.k}/q'_{.k}$, so that the weighted row distribution becomes $W_{.k}$. As a result of this second weighting, the column distribution is no longer $W_{h.}$, but has been changed to $\Sigma_k q'_{hk} W_{.k}/q'_{.k} = \Sigma q''_{hk} = q''_{h.}$. The procedure then returns to the first step applied to the q''_{hk} values, and continues to iterate until the process converges. A variant is to continue the process for a small number of cycles, say from one to four, and then to terminate it regardless of whether it has converged to a specified level. The resulting distribution, which will be denoted by $\tilde{w}_{hk}$, is then used in the estimation of the population mean as $y_q = \Sigma\Sigma\tilde{w}_{hk}\bar{y}_{rhk}$. This estimator may be alternatively expressed as $y_q = \Sigma\Sigma\Sigma w_{hkj} y_{rhkj}/\Sigma\Sigma\Sigma w_{hkj}$, where $w_{hkj} \propto \tilde{w}_{hk}/r_{hk}$ and r_{hk} is the sample size in cell (h,k).

The assumption underlying raking ratio estimation is that the pattern of response rates across all the cells of the table conforms to the following logit model: $\ln[\bar{R}_{hk}/(1 - \bar{R}_{hk})] = a_h + b_k$. This may be loosely illustrated by the application of the raking algorithm to the respondents' expected joint distribution across the cells, W_{rhk}, under the special case of the logit model when $\bar{R}_{hk} \propto \bar{R}_{h.}\bar{R}_{.k}$. The initial two steps in the algorithm are then first to weight the (h,k) cell by $W_{h.}/W_{rh.} = \bar{R}/\bar{R}_{h.}$, and then by $W_{.k}/\Sigma_h(W_{rhk}\bar{R}/\bar{R}_{h.})$. Since $W_{rhk} = W_{hk}\bar{R}_{hk}/\bar{R}$, after these two steps the weighted value for cell (h,k) is

$$W_{rhk}(\bar{R}/\bar{R}_{h.})[W_{.k}/(\Sigma_h W_{rhk}\bar{R}/R_{h.})] = (W_{hk}\bar{R}_{hk}/\bar{R})(\bar{R}/\bar{R}_{h.}) \cdot [W_{.k}/(\Sigma_h W_{hk}\bar{R}_{hk}/\bar{R}_{h.})]$$

$$= W_{hk}(\bar{R}_{hk}/\bar{R}_{h.})[W_{.k}/(\Sigma_h W_{hk}\bar{R}_{hk}/\bar{R}_{h.})].$$

Under the model $\bar{R}_{hk} \propto \bar{R}_{h.}\bar{R}_{.k}$, it follows that the weighted value converges immediately to $\tilde{W}_{hk} = W_{hk}$, as required, and no further iterations are needed.

The bias of the raking ratio estimator of the population mean is given by

$$B(\bar{y}_q) = E_1 E_2(\Sigma\Sigma w_{hk}\bar{y}_{rhk}) - \bar{Y}$$
$$= \Sigma\Sigma E_1(\tilde{w}_{hk})\bar{Y}_{rhk} - \bar{Y}$$

Letting $E_1(\tilde{w}_{hk}) = \tilde{W}_{hk}$, and employing (3.2.1),

$$B(\bar{y}_q) = \Sigma\Sigma\tilde{W}_{hk}\bar{Y}_{rhk} - \Sigma\Sigma W_{hk}\bar{Y}_{rhk} + \Sigma\Sigma W_{hk}\bar{M}_{hk}(\bar{Y}_{rhk} - \bar{Y}_{mhk})$$

i.e. $$B(\bar{y}_q) = \Sigma\Sigma W_{hk}\bar{M}_{hk}(\bar{Y}_{rhk} - \bar{Y}_{mhk}) + \Sigma\Sigma(\tilde{W}_{hk} - W_{hk})\bar{Y}_{rhk}. \quad (3.4.1)$$

The form of the bias in (3.4.1) brings out its dependence on two factors: the magnitude of the first term, the bias of a population or a sample weighting estimator applied to the cells of the crossclassification, depends on the differences between respondent and nonrespondent means in the cells; the second term depends on the closeness of the model estimates of the population proportions in the cells, $\tilde{W}_{hk}$, and the true population proportions in the cells, W_{hk}. Further insight into the second term can be gained from expressing $\bar{Y}_{rhk}$ as

$$\bar{Y}_{rhk} = \bar{Y}_r + (\bar{Y}_{rh.} - \bar{Y}_r) + (\bar{Y}_{r.k} - \bar{Y}_r)$$
$$+ (\bar{Y}_{rhk} - \bar{Y}_{rh.} - \bar{Y}_{r.k} + \bar{Y}_r).$$

Then, noting that $\Sigma_h\tilde{W}_{hk} = W_{.k}$ and $\Sigma_k\tilde{W}_{hk} = W_{h.}$ with the raking procedure, the second term in (3.4.1) reduces to

$$\Sigma(\tilde{W}_{hk} - W_{hk})(\bar{Y}_{rhk} - \bar{Y}_{rh.} - \bar{Y}_{r.k} + \bar{Y}). \quad (3.4.2)$$

Thus, even if $\tilde{W}_{hk}$ and W_{hk} were different, this term would be zero if there were no interaction in the $\bar{Y}_{rhk}$ for the classifications involved.

The variance of $\bar{y}_q$ conditional on the r_{hk}'s is obtained straightforwardly as

$$v_2(\bar{y}_q) = \Sigma\Sigma\tilde{w}_{hk}^2(1 - f_{hk})s_{rhk}^2/r_{hk}$$

where $f_{hk} = 1 - (r_{hk}/R_{hk})$. The unconditional variance is complex even for SRS (see, for instance, Brackstone and Rao, 1980).

Given the joint distribution of the characteristics to be included in a population adjustment procedure, the analyst has the choice between separate cell-by-cell adjustments and raking. An aspect of that choice includes the possibility that some cells may be collapsed with the population adjustment procedure to give greater stability of weights. The possibility also exists of using raking for some cells, for instance those with few cases, and using population weighting adjustments for the remainder. The choice involves consideration of a trade-off between bias and variance, and no general rules can be laid down. The best choice depends on the circumstances and, in particular, on the suitability of the raking model for the data being analyzed.

Although primarily described here in terms of marginal population weighting adjustments, raking can also be applied with sample weighting adjustments: the marginal sample distributions of the defining characteristics are determined for the total sample of respondents and nonrespondents combined, i.e. $n_{i.}/n$ and $n_{.j}/n$, and then the raking algorithm is applied as above.

Raking can be employed for marginal post-stratification adjustments as well as for nonresponse weighting adjustments. References on raking include Arora and Brackstone (1977a,b), Bankier (1978), Oh and Scheuren (1978a,b), Brackstone and Rao (1980), and Konijn (1981).

3.5 Weighting Using Response Probabilities

Several different approaches have been suggested for employing information about the difficulty of securing responses from respondents in nonresponse weighting adjustments. The well-known Politz-Simmons procedure is one example (Politz and Simmons, 1949, 1950). The essence of this procedure is to group respondents into classes according to their response probabilities, and then to weight the classes by the inverses of these probabilities. The procedure was designed to obviate the need for more than one call on each sample unit. During the interview respondents are asked on how many of the previous five evenings they were home at about that time. Those at home every evening were then assigned response probabilities of 1, those home on 4 of the 5 evenings were assigned probabilities of 5/6, etc., and those home on none of the previous 5 evenings were assigned probabilities of 1/6 (the evening they were interviewed being their one chance in six of being interviewed). On the grounds that only a fraction t/6 of those in t/6 evenings will be found at home on one call, this class is then weighted by the factor 6/t. Note that the procedure fails to account for those in none of the six evenings, since none of them are interviewed; they should, however, constitute a small proportion of the total sample. It should also be observed that the Politz-Simmons procedure is designed to deal with not-at-homes, not refusals. However, as shown in Chapter 1, the main unit nonresponse problem facing the 1978 Research Panel was refusals.

The Politz-Simmons procedure requires respondents to provide the information on their own response probabilities. An extension to this procedure avoids the need to collect this information. It involves making call-backs on nonrespondents at earlier calls, modelling the response probabilities at successive calls, and estimating the parameters of the model from the observed pattern of responses at successive calls. This approach can be carried out within different classes of the sample to allow for differing parameters across the classes.

Drew and Fuller (1980, 1981) describe one application of this approach. Suppose that data can be assumed to be missing at random within classes of a characteristic, but at different rates between classes, so that the characteristic would be useful for weighting adjustments. If the proportion of the population in each class, W_h, was known, a population weighting adjustment could be used, estimating the population mean by $\bar{y}_p = \Sigma W_h \bar{y}_h$. If the characteristic were known for nonrespondents as well as for respondents in the sample, the sample weighting adjustment could be used, estimating the population mean by $\bar{y}_s = \Sigma w_h \bar{y}_h$. If neither of these sets of information is available, the Drew and Fuller model provides means of estimating W_h, $\hat{W}_h$, which can then be used in a weighting adjustment as $\Sigma \hat{W}_h \bar{y}_h$. The Drew and Fuller model assumes that each unit in class h has the same response probability, q_h, and that q_h remains the same for each successive call; it also allows for a proportion of hard-core nonrespondents, γ, which is assumed constant across classes. Given these assumptions, the number of responses obtained at each successive call follows a multinomial distribution. The parameters q_h, γ and W_h can then be estimated by fitting the model to the observed data of the number of responses at each call for each class of the characteristic.

Thomsen and Siring (1979) adopt a similar approach to Drew and Fuller, but with a more complex model. They assume that within a class a unit has a probability of p_h of responding at the first call, and of $\Delta_h p_h$ at subsequent calls; they expect Δ_h to exceed 1 because interviewers can make appointments, etc., for second and subsequent calls. They also allow for a proportion of hard-core nonrespondents. They give an application in which the weighting class characteristic is itself the variable of interest: they estimate the mean number of live births per woman, and form classes of women with 0, 1, ..., 6 or more births. This can be viewed as an interesting special case of the general procedure with $\bar{y}_h = (h - 1)$.

Chapman (1976) describes the use of regression analysis to estimate response probabilities. A response variable (1 for respondents, 0 for nonrespondents) is regressed on a set of variables available for both respondents and nonrespondents, and the predicted values of the response variable are computed for the respondents. These predicted values estimate the response probability for units with the same values on the independent variables, and their inverses are then used as weights for the respondents. As described, this approach makes strong assumptions about the linearity and additivity of the effects of the independent variables on the response probabilities. It would, however, be possible to modify the model to take account of some nonlinearities and interactions.

Another type of approach using information on the differential difficulty of securing responses from different respondents is to assume that difficulty is correlated with the survey variables. This leads to two possible procedures for handling nonresponse.

One is to assume that the nonrespondents have similar values on the survey variables to those respondents who were hardest to interview - commonly taken as those who were interviewed only after a number of call-backs. Bartholomew (1961), for instance, has suggested making only two calls in a survey, and weighting up the responses obtained at the second call to compensate for the nonrespondents. This kind of procedure would be effective in reducing bias if the basic assumption holds. However this is often questionable; in particular the nonrespondents comprise two groups, not-at-homes and refusals, and it is not obvious that the late respondents can properly represent the refusals. Based on an intensive follow up study of nonrespondents in the Current Population Survey, Palmer and Jones (1966) and Palmer (1967) found no support for the assumption. Another problem with this procedure is that there is typically only a small number of late respondents, who then need to be assigned sizeable weights if they are to compensate for the nonrespondents. This factor considerably increases the variances of the survey estimates.

The second possible approach for handling nonresponse taking account of a relationship between the survey variables and difficulty in securing responses is to plot the means of the survey variables against difficulty, probably measured by number of calls needed. Any trend observed could then be extrapolated to estimate the mean for the nonrespondents. This procedure assumes that nonrespondents follow along the trend after the last respondents, often a dubious assumption for the reasons given above. Chapman (1976) reports that in his investigation of such trends for many items in the Health and Nutrition Examination Survey there was no general pattern to the trends observed, and in most cases the trends were not clear enough to be used for projecting the means for nonrespondents.

3.6 Concluding Remarks

So far the important issue of how to choose the weighting classes has been mentioned only in passing. The first consideration in making this choice is to ascertain what variables are available for use, that is what variables satisfy the data requirements for a weighting procedure. Population weighting adjustments require knowledge of the population distribution of the variable and information on the variable for survey respondents. Sample weighting adjustments require information on the variable to be available for both respondents and nonrespondents.

Among the potential variables for use in forming weighting classes, the ones that are most effective in reducing nonresponse bias are those that are highly correlated both with the survey variables and with the (0 - 1) response variable (see the discussion in Section 3.2). Since a survey is concerned with numerous survey variables, which may have markedly different correlations with any potential weighting class variable, this aspect of the choice may be confusing unless a few closely-related survey variables can be identified as the main concern. The correlation between a potential weighting class variable and the response variable is a single criterion that applies no matter which survey variable is being analyzed. In consequence considerable importance may be attached to this correlation in making the choice.

Weighting classes can be defined in terms of a combination of variables, and this is indeed common practice. The classes may be simply defined as the cells of a crossclassification of the variables, but there is no need for this balanced approach. Instead some variables may be used to define weighting classes in one part of the sample while others can be used in other parts. Also, if the crossclassification approach is used, some cells may be collapsed to form weighting classes of adequate size. The position corresponds to that involved in choosing strata for stratified sampling, where the strata may be defined in any way providing the required information is available.

The number of variables that can be used in forming weighting classes is limited by the sizes of the resulting classes. If the sample sizes of respondents in the classes are too small, the effect of the resulting unstable weighting will be to substantially inflate the variances of the survey estimates, more than would be compensated for by any bias reduction. Rules of a minimum of 20 or 25 respondents per class have been used (Chapman, 1979). Another condition sometimes imposed is that any class that requires a weighting factor n_h/r_h greater than a specified level should be combined with an adjacent class; this condition also serves to limit the number of weighting classes. In the Current Population Survey (CPS), for instance, adjacent weighting classes are combined if at least one of the classes requires a weighting factor greater than 2.0. In addition, if after combining two adjacent classes in the CPS the weighting factor exceeds 3.0, it is trimmed down to 3.0 (Bailar et al., 1978; Brooks and Bailar, 1978). These procedures in effect accept some potential loss in bias reduction in order to avoid the increase in variance associated with large variability in weights.

For convenience of exposition, this chapter has discussed population and sampling weighting adjustments separately, but it should be noted that they can be - and often are - applied together. First, sample weighting adjustments may be employed to bring the achieved sample into line with the total sample, primarily in terms of the sample design variables, and then population weighting adjustments can be applied, partly to handle noncoverage and total nonresponse and partly to serve as post-stratification.

4. Imputation

4.1 Introduction

As noted in Chapter 1, most surveys face problems of item nonresponse as well as unit nonresponse. Item nonresponse occurs when no answer is recorded for a question that a respondent should have answered, and when an answer is eliminated by an edit check because it is inconsistent with other answers. Imputation is the process of assigning values for the missing values to produce a complete data set.

There are several benefits to imputation. In the first place, like weighting adjustments, the aim of imputation is to reduce biases in survey estimates. Secondly, imputation makes analysis easier and the results simpler to present. Complex algorithms to estimate population parameters in the presence of missing data are not required, and hence much processing time is saved. There is neither a need to determine the different sets of cases with missing data that have to be deleted from different analyses, nor a need to provide details of the extent and treatment of missing data with each set of results. Thirdly, the results obtained from different analyses are bound to be consistent with one another, a feature which need not apply to results of analyses from an incomplete data set. In particular, the practice of deleting cases with missing values on any of the classifications involved in forming a tabulation from an incomplete data set can give rise to inconsistencies in marginal distributions between tabulations having some classifications in common.

On the other hand, imputation has its dangers. There can be no guarantee that the results obtained after imputation will be less biased than those based on the incomplete data set, and indeed the biases could be greater. It all depends on the suitability of the assumptions built into the imputation procedures used. Even if the biases of univariate statistics are reduced, the relationship between variables may be distorted (see the companion report by Santos, 1981a, which examines this issue). Rockwell (1975) shows, for instance, how this point applies to relationships between the variables subject to imputation and the variables employed in making the imputations. Another danger with imputation is that researchers may falsely treat the completed data set for n respondents as if it were from a straightforward sample of size n. Imputation merely fabricates data; it does not increase the amount of information available, and indeed the use of a random subsampling of donors serves to reduce, not to increase, the effective sample size. It is sometimes argued that little risk is involved when imputation is needed for only a small fraction of cases; however, even in this situation, some subgroups of the sample may have sizeable proportions of missing data, and hence their results may significantly depend on the imputed values. (See Banister, 1980, for some concerns about the widespread use of imputation with census data.)

Undoubtedly, when used, imputation should be applied cautiously and analysts of the completed data set should be fully warned of the potential dangers created by the imputation. At a minimum, the imputed values should be flagged, so that the careful analyst can assess the effect that imputations may have on the analysis. I.G. Sande (1979a) has suggested a number of additional aspects of the imputation process that need to be audited in order to keep track of the impact of imputation; these include such aspects as the number of times a particular record is used as a donor of imputed values in a particular stage, the number of attempts made to achieve successful imputations for a particular record with missing data, and information on which donors contribute which fields to which recipients. A second reason for flagging imputed values is to enable an analyst to replace them with his own estimates, or to treat the missing data otherwise in his analyses. Imputation is intended to be a general-purpose strategy, good for many forms of analysis but not necessarily optimum for any one (in this sense it is like Tukey's description of the jackknife as a technique of variance estimation). An analyst who has the resources may prefer to ignore the imputed values and instead treat the incomplete data set in a way that is tailor-made to his particular problem.

Given that imputed values are to be substituted for missing responses, there are a variety of methods in which the imputed values may be determined. A selection of these methods is reviewed in the next section, and the results of a simulation study applying some of them to a data set derived from the April 1978 wave of the 1978 ISDP Research Panel are presented in Section 4.3.

Many of the imputation methods divide the sample into imputation classes corresponding to the weighting classes discussed in Chapter 3. The range of choice for imputation classes is far greater than that for weighting classes because many more data are available for the units with item nonresponses. A discussion of methods for choosing imputation classes is given in Section 4.4. That section also discusses a number of other issues that occur with imputation. These include issues arising from using respondents as donors on several occasions, the need to ensure that imputed values are consistent with the unit's other responses, and the considerations involved when a record has several missing items.

4.2 Imputation Methods

The purpose of this section is to review a range of methods of imputing for missing values. Since there are several variants possible for many of the basic methods, the review does not attempt an exhaustive coverage but rather aims to describe some of the main approaches. For simplicity, the descriptions will be given in terms of missing responses for a single item; the complexities introduced by records having several missing values are taken up in Section 4.3.

The development of imputation methods is of recent origin, and in consequence no universally accepted terminology has yet been established. It will therefore be useful to start by defining some of the terms employed in the following discussion. First, most of the methods start with the division of the sample into classes, which we will call <u>imputation classes</u>. Imputation classes are equivalent to the weighting classes discussed in the previous chapter. The variables used to define the imputation classes, or otherwise used in determining the imputed values, will be termed <u>control</u> or <u>matching variables</u>. Many methods assign the value from a record with a response on the item in question to a record with a missing value on that item. These records will be called the donor and recipient respectively. Any method that takes the donors from within the same sample as the recipients is often called a "hot-deck" method; however, for the present discussion the term hot-deck is reserved for a particular method using this approach, a sequential method as used, for instance, with the labor force items in the Current Population Survey (Brooks and Bailar, 1978).

It is useful to make a distinction between qualitative and quantitative control variables. The formation of imputation classes in terms of qualitative control variables is in principle straightforward, but when quantitative variables are used, they first need to be categorized into a number of classes, necessarily involving some loss of information. The loss of information is in fact relatively slight providing a reasonable number of well-chosen classes is used, say 5 or more (see, for instance, Cochran, 1968; Aigner *et al*., 1975; Kish and Anderson, 1978). However, the categorization can be avoided by the use of alternative imputation methods not involving imputation classes, such as methods based on minimizing a distance function between donor and recipient in terms of the control variables or methods using regression for determining the imputed values. These methods are discussed in (g) and (h) below.

We will now turn to descriptions of a selection of methods available for making imputations for missing values. Some of these methods have in fact already been introduced in Chapter 2, where they have been considered in relation to the single imputation class comprising the sample taken as a whole. These methods will be repeated here briefly for the sake of completeness.

(a) Deductive imputation. In its pure form this method of imputation can be applied in situations in which a missing response can be deduced with certainty, or with high probability, from other information on the record. Thus, for instance, if a respondent's sex is missing, but the person has a female name and is known to be married to a male, the sex of the respondent may be deduced to be female. If a record should contain a series of amounts together with a total but one of the quantities is missing, the missing value can be deduced from the others. This method essentially depends on some redundancy in the information collected so that edit constraints can determine the missing values on some items.

Deductive imputation is the special case of other imputation methods in which the control variables determine the missing values with negligible error; it merges into other types of imputation as the amount of error increases. This may be illustrated by another example, taken from the 1978 ISDP Research Panel. Wage earners were asked to report their quarterly earnings from their records (E) and were also asked to give their hourly rates of pay (P), their usual numbers of hours worked (H) and the numbers of weeks worked in the quarter (W). In cases where they were unable or unwilling to consult their records, their missing earnings could be estimated by deductive imputation to be $E' = (P \times H \times W)$. This deductive process is, however, an imperfect one, as is evidenced by the discrepancies found between E and E' for those wage earners who provided both sets of data (discussed further below). Deductive imputation ignores these discrepancies; if they are too substantial to be ignored, other imputation methods are needed to deal with them.

(b) Cold-deck imputation. The cold-deck imputation method was an early form of computerized imputation which has now been superseded by hot-deck methods. It is mentioned here mainly for the sake of completeness.

The essence of the method involves the specification of imputation classes, often defined as the cells of a cross-classification of the control variables, and then, for each class, the provision of a set of values for the responses to the item in question. These values come from some past data set for the same population; with a periodic survey, they may be obtained from the respondents within the imputation classes on the previous round of the survey. Once this cold-deck has been established, the current survey records are considered in turn. If a record has a missing response for the item, it is assigned a value from the cold-deck. Thus a complete data set is readily produced.

The evident objection to the cold-deck method is the potential lack of comparability between the past values inserted into the cold deck and the present survey responses. This lack of comparability might arise both because of obsolescence and because of differences in procedures and definitions used. The obvious solution to this problem is to employ a method based on current values rather than past values for the imputation process. This leads to the hot-deck method of imputation as described in (d) below.

(c) Mean-value imputation. The mean-value imputation procedure consists of dividing the sample into imputation classes, computing the respondent mean within each class, and assigning the class mean value to all records in the class with missing data for the item. This procedure has already been discussed in Chapter 2, where it was shown to be a good procedure for the estimation of the item mean. However its distortion of the distribution of values makes it unsuitable for many other forms of analysis. These conclusions also apply to the more general situation of several imputation classes now being considered.

(d) Hot-deck imputation. Contrary to some other usages, the term 'hot-deck' imputation will be used here only for the sequential kind of procedure used with the labor force items in the Current Population Survey (CPS), sometimes known as the traditional hot-deck procedure. This procedure begins with the specification of imputation classes, and with the assignment of a single cold-deck value for each class to provide a starting point for the process. The records of the current survey are then treated sequentially. If a record has a response for that item, that response replaces the value stored for its imputation class at that point; if the record has a missing response it is assigned the value stored for its imputation class.

This hot-deck procedure was developed to remedy the disadvantage of the cold-deck by using the survey's respondents as donors, while at the same time enabling the imputations to be made from a single pass through the data file. Another argument that is made in favor of the sequential nature of the procedure is that the imputation of a value to a recipient from the last donor in his imputation class serves as an additional degree of matching, which will be beneficial if the file is ordered in a way that creates positive autocorrelation.

The hot-deck procedure has, however, some serious disadvantages. One is its dependence on the order of the file without a probability mechanism: this feature makes it impossible to develop a model-free theoretical evaluation of the procedure. Secondly, the procedure may easily give rise to multiple uses of donors. This event occurs whenever within a given imputation class a record with missing data is followed by one or more other records with missing data. All these records are assigned the value then being stored for the imputation class, namely the value of the last respondent. Thirdly, the number of imputation classes has to be limited to ensure that one or more donors will be found for each class containing a record with a missing value.

Useful discussions of the hot-deck procedure are provided by Bailar *et al.* (1978), Bailar and Bailar (1978, 1979), Ford (1980), Oh and Scheuren (1980), Oh *et al.* (1980) and I.G. Sande (1979a,b).

(e) *Random imputation*. Random imputation is used here to denote a range of procedures in which donors are selected by some form of probability sampling within an imputation class. A number of these procedures have already been described in Chapter 2 for a single imputation class. When there is more than one imputation class, the first step is to identify the n_h sampled units in imputation class h, and to separate them into r_h potential donors (units with responses to the item) and m_h recipients (units with missing responses to the item). Then one of the procedures described in Chapter 2 can be applied within each imputation class.

To recap, there are two main features to be determined in arriving at a probability sampling scheme for donors. One is the sample size and the other is the sample design. The natural choice for sample size is m_h, allocating to each recipient the value from one donor. Given this sample size, a simple sample design would be to select one donor at random from the r_h potential donors for each recipient, that is to sample donors by unrestricted sampling. This scheme of sampling with replacement gives rise, however, to an undesirable variability in the use of donors. This variability can be controlled by simple random sampling, sampling without replacement, using the procedure described in Chapter 2 of taking each donor k_h times and a SRS of t_h donors for an extra occasion, where $m_h = k_h r_h + t_h$ and k_h is an integer (if, as often applies, $m_h < r_h$, $k_h = 0$). Providing $m_h > 1$, the sample design can be further improved by the use of stratification (in particular stratifying potential donors by their item responses) or by systematic sampling from a list of potential donors ordered by their item responses. The maximum number of strata - or implicit strata with systematic sampling - that can be employed is t_h.

The donor sample size for random imputation (or indeed for other imputation procedures) need not be set at $m_h = k_h r_h + t_h$, with $(r_h - t_h)$ respondents acting as donors to k_h recipients each and t_h respondents acting as donors to $(k_h + 1)$ recipients. A more general approach is to take a sample of $c_h t_h$ respondents to serve as donors on the additional occasion, with $c_h t_h \leq r_h$ and c_h a positive integer. Then $k_h r_h$ of the m_h recipients are given values from a single donor, and the other t_h are given values from c_h donors each. The multiple donor cases may be handled by splitting the records of each of the recipients involved into c_h parts, allocating one of the c_h donors' values to each part, and assigning a weight of $1/c_h$ to each part. (Note that if several parts for one recipient happen to receive the same value, they can be recombined with their weights added.) The advantage of using this type of multiple imputation is the reduction in variance increase associated with using a sample of respondents as donors for an additional occasion (see equation 2.3.1.13).

A variant of the above procedure that has been proposed in the literature is to select a sample of donors for each recipient and to assign to each recipient the mean value for his donor sample. This procedure has the same effect for estimating the mean and total of an item as the weight-splitting procedure described above. The spread of the distribution of imputed values obtained from this procedure is, however, reduced since it is a distribution of means, and hence the overall distribution of respondents' values and imputed values will be distorted. The limiting case of this procedure is the mean-value imputation procedure described in (c) above.

Its disadvantages are the same as those of mean-value imputation, only in a reduced form.

Consider now the bias and variance of a sample mean $\bar{y}_k$ when random imputation has been used to fill in the missing values. We assume that the original sample is an epsem one and that $\bar{y}_k = \Sigma\Sigma y_{hi}/n = (\Sigma\Sigma y_{rhi} + \Sigma\Sigma y_{mhi})/n$, where y_{rhi} is the y-value for responding unit i in imputation class h and y_{mhi} is the imputed y-value for a nonresponding unit i in imputation class h. The expectation of $\bar{y}_k$ may be expressed as $E_1E_2(\bar{y}_k)$ where E_2 is the expectation conditional on n_h, r_h and the responding units' values y_{rhi}, and E_1 is expectation over variations in these values. Then

$$E(\bar{y}_k) = E_1E_2(\Sigma\Sigma y_{rhi} + \Sigma\Sigma y_{mhi})/n.$$

For any epsem sampling scheme for choosing donors, $E_2(y_{mhi}) = \bar{y}_{rh}$, so that

$$\begin{aligned} E(\bar{y}_k) &= E_1(\Sigma r_h\bar{y}_{rh} + \Sigma m_h\bar{y}_{rh})/n \\ &= E_1(\Sigma n_h\bar{y}_h/n) = E_1(\bar{y}_s) \end{aligned}$$

where $\bar{y}_s$ is the mean with sample weighting adjustments. The expectation of $\bar{y}_s$, as obtained in Section 3.3, is $\Sigma W_h\bar{Y}_{rh}$, so that

$$E(\bar{y}_k) = \Sigma W_h\bar{Y}_{rh}.$$

The bias of $\bar{y}_k$ is the same as that of $\bar{y}_s$ and $\bar{y}_p$, the mean with population weighting adjustments. See Section 3.2 for a discussion of this bias.

The unconditional variance of $\bar{y}_k$ may be obtained from

$$\begin{aligned} V^*(\bar{y}_k) &= E_1V_2(\bar{y}_k) + V_1E_2(\bar{y}_k) \\ &= E_1V_2(\bar{y}_k) + V_1(\bar{y}_s). \qquad (4.2.1) \end{aligned}$$

The second term in (4.2.1) is the variance of the mean with simple weighting adjustments, $V^*(\bar{y}_s)$. This variance is derived for an initial SRS of n units in Section 3.3, with $V^*(\bar{y}_s)$ being given by (3.3.6). The first term in (4.2.1) is the additional variance arising from the sampling of respondents to be used as donors: the conditional variance $V_2(\bar{y}_k)$ for various sampling schemes has been considered in Section 2.3.1. One such scheme, for example, is the SRS imputation scheme (within each class). In this case

$$V_2(\bar{y}_k) = \Sigma t_h[1 - (t_h/r_h)]s_{rh}^2/n^2 \qquad (4.2.2)$$

as may be readily obtained as a generalization of (2.3.1.5). As noted above, this variance may be reduced by employing a stratified or systematic sampling scheme and by increasing the sample size.

A reasonable estimator of $V^*(\bar{y}_k)$ may be readily derived as follows. First, as shown in Section 3.3, $v(\bar{y}_s)$ in equation (3.3.8) serves as a good estimator of $V^*(\bar{y}_s)$. Secondly, $V_2(\bar{y}_k)$ can be computed directly from (4.2.2). Hence

$$v(\bar{y}_k) = V_2(\bar{y}_k) + v(\bar{y}_s)$$

may be used to estimate $V^*(\bar{y}_k)$.

(f) Flexible matching imputation. The term flexible matching imputation has been coined here to denote the modified hot-deck procedure now being used by the Bureau of the Census for the March Income Supplement of the CPS. The essence of this sophisticated procedure is a sorting and matching of potential donors and recipients on a sizeable number of control variables. The matching is done on a hierarchical basis, in the sense that if no donor can be found to match a recipient on all the control variables, some of the control variables considered less important are dropped or their detail of classification is reduced, and the match is made at a lower level. Three hierarchical levels are defined for this purpose, with the lowest level being such that a matching donor can always be found. This flexible matching procedure enables far closer matches to be secured for many recipients than does the conventional hot-deck procedure. The procedure also avoids the multiple use of donors in classes where the number of donors is not less than the number of recipients. Further details of the implementation and evaluation of the procedure with the March

Income Supplement of the CPS are given by Coder (1978) and Welniak and Coder (1980).

(g) Distance function matching. Matching on quantitative control variables may be carried out by categorizing the variables and forming imputation classes. This approach can, however, lead to some less desirable matches: for instance, a recipient with a value on the control variable close to the upper bound of a category may be matched with a donor close to the category's lower bound, whereas he might have been better matched with a donor near the lower bound of the next higher category. The step function effect created by fixing arbitrary boundaries for categories can be avoided by the use of some form of distance function to measure the "closeness" of the match.

When a single quantitative control variable is involved, the merged set of recipients and potential donors can be ordered by it, and the "nearest" neighboring donor to a recipient could be taken. "Nearest" can be defined in terms of the absolute difference between recipient's and donor's values on the control variable, or alternatively in terms of the absolute difference on some transformation of the control variable; for example, using a logarithmic transformation in effect measures "nearest" in terms of proportionate differences. The distance function can be constructed to include a component to reduce the multiple use of donors. Colledge et al. (1978), for instance, define the distance between a potential donor and a recipient to be $D(1 + pd)$, where D is the basic distance, d is the number of times the donor has already been used, and p is a proportional penalty for each usage (they give $p = 0.02$ as an example). A variation of the distance function procedure is to assign the recipient the average value for a set of nearest neighbors, a simple example of which is to average the values of the neighboring donors on either side of the recipient (Ford, 1976). As with other averaging procedures, this procedure suffers the disadvantage of distorting distributions.

When several control variables are used, but only one is quantitative, the above procedure may be applied for the quantitative variable within imputation classes constructed from the qualitative variables. When there is more than one quantitative variable, the distance function can be generalized. In this case, attention needs to be paid to the metrics and distributions of the variables. Transformations of the variables are almost certainly needed. In particular, distances in the tails of a distribution where observations are sparse are likely to dominate the overall distance function unless a suitable transformation is made. One way of eliminating this effect is to transform all the quantitative control variables to ranks, and define distances in terms of ranks (suggested by G. Sande), or equivalently to transform the control variables to a uniform distribution. One form of distance function between recipient r and potential donor d might then be

$$\mathrm{Sup}_k w_k \, |R_{rk} - R_{dk}|$$

where R_{rk} and R_{dk} are the ranks of the recipient and donor on variable k, and w_k is a weight representing the relative importance of variable k in the distance function (see I.G. Sande, 1979b). Vacek and Ashikaga (1980) suggest an approach based on the Mahalanobis distance. Qualitative variables can also be incorporated into a distance function, as an alternative to controlling for them by means of imputation classes.

(h) Regression imputation. With this procedure the item for which imputations are needed (y) is regressed on the control variables ($x_1, x_2, \ldots, x_p$) for the units providing a response on y. The control variables may be quantitative or qualitative, the latter being incorporated into the regression model by means of dummy variables. If the y-variable is qualitative, log-linear or logistic models may be employed. The missing values may then be imputed in one of two basic ways: one is to use the predicted value from the model given the values on the control variables for the record with a missing item response; the other is to use this predicted value plus some type of randomly chosen residual.

It will be helpful at this point to compare regression imputation with imputation procedures based on imputation classes. For this purpose, we will consider a situation with two quantitative control variables, each with a few discrete scores so that no categorization is needed for forming imputation classes (one variable, for instance, might be number of children in the household, ranging from, say, 1 to 4). The model underlying procedures based on imputation classes is

$$y_{ijk} = \bar{Y}_{ij} + e_{ijk} \tag{4.2.3}$$

where y_{ijk} is the value for unit k in the class represented by the intersection of the i-th level of control variable x_1 and the j-th level of control variable x_2, $\bar{Y}_{ij}$ is the population class mean and e_{ijk} is a residual representing the departure of the values of unit k from the class mean with $E(e_{ijk}|ij) = 0$. The distribution of errors is not specified, and it may differ between classes. With this model, the predicted value of y_{ijk} is $y_{ijk} = \bar{Y}_{ij} = \bar{y}_{ij}$, that is the mean-value imputation as discussed in (c) above. To retain the distributional properties of the y_{ijk} it is necessary to include the residual term, i.e. to estimate $(\bar{Y}_{ij} + e_{ijk})$. A natural approach to estimating e_{ijk} is to select a potential donor at random, say donor k', from class ij and to compute $\hat{e}_{ijk} = y_{ijk'} - \hat{\bar{Y}}_{ij}$. From the model this estimator is

$$\hat{e}_{ijk} = \bar{Y}_{ij} + e_{ijk'} - \bar{Y}_{ij} - \bar{e}_{ij} = e_{ijk'} - \bar{e}_{ij}$$

and hence is not a good estimator of e_{ijk}. However,

$$\hat{y}^*_{ijk} = \hat{\bar{Y}}_{ij} + \hat{e}_{ijk} = \bar{Y}_{ij} + \bar{e}_{ij} + e_{ijk'} - \bar{e}_{ij} = \bar{Y}_{ij} + e_{ijk'}$$

does estimate y_{ijk}. Thus $\hat{y}^*_{ijk}$ is a suitable value to impute for y_{ijk}. In practice, y^*_{ijk} need not be computed from $\hat{\bar{Y}}_{ij} + \hat{e}_{ijk'}$, because this quantity is simply equal to $y_{ijk'}$. In other words the recipient can be simply assigned the value of a randomly-chosen donor. This is random imputation, as discussed in (e) above, within class ij.

The model (4.2.3) can be reexpressed as

$$y_{ijk} = \bar{Y} + a_i + b_j + c_{ij} + e_{ijk} \tag{4.2.4}$$

where $a_i = (\bar{Y}_i - \bar{Y})$ and $b_j = (\bar{Y}_j - \bar{Y})$ are the main effects and $c_{ij} = (\bar{Y}_{ij} - \bar{Y}_i - \bar{Y}_j + \bar{Y})$ is the interaction effect of control variables i and j in the usual analysis of variance representation. Since no assumption is made about the distribution of the residuals e_{ijk}, this formulation makes clear the lack of assumptions made by the imputation class model. However many control variables are used in defining classes, all interactions of all levels are included in the model. The only assumption involved is the necessary one that recipients are missing at random.

The cost of the generality of the imputation class model is that less good estimators may be obtained than would come from a model in which more assumptions were made - providing those assumptions held. The restrictions imposed on the model by making extra assumptions may also permit the model to be extended in other ways, in particular by including more control variables in it.

Several types of assumptions could be made. One, for instance, would be to assume that the distribution of residuals is the same in all classes, perhaps even that it is a normal distribution; we will return to this later when we consider regression. Another is to assume that there are no interaction effects, for instance $c_{ij} = 0$ in (4.2.4); in this case (4.2.4) can be viewed as a dummy variable regression. (Alternatively, one could specify that just certain interactions may be present.) In passing it is worth noting the relationship between the no-interaction model and raking ratio estimation discussed in Section 3.4.

In addition to the assumption of no interactions (additivity), linear relationships between the dependent variable and the control variables might be assumed. The model then becomes a multiple regression model

$$y_k = B_0 + B_1x_{1k} + B_2x_{2k} + e_k$$

where the single subscript k now runs over the whole sample. Furthermore, the residuals may be assumed to be homoscedastic and even normally distributed.

With this background, we now return to the issue of how to impute for missing values with regression imputation. The use of the predicted value

$$\hat{y}_k = \hat{B}_0 + \Sigma\hat{B}_ix_{ik}$$

corresponds to the mean-value imputation in the restricted model, and hence has the same undesirable distributional properties. A good case therefore exists for including an estimated residual. There are various ways in which this could be done depending on the assumptions made about the residuals. The following are some of the more obvious possibilities:

(i) Assume that the errors are homoscedastic and normally distributed, $N(0,\sigma_e^2)$. Then σ_e^2 could be estimated by the residual variance from the regression, s_e^2, and the residual for a recipient could be chosen at random from $N(0,s_e^2)$.

(ii) Assume that the errors are heteroscedastic and normally distributed, with σ_{ej}^2 being the residual variance in some group j. Estimate the σ_{ej}^2 by s_{ej}^2, and choose a residual for a recipient in group j from $N(0,s_{ej}^2)$.

(iii) Assume that the residuals all come from the same, unspecified, distribution. Then estimate y_k by $\hat{y}_k + \hat{e}_{k'}$, where $\hat{e}_{k'}$ is the estimated residual for a random-chosen donor.

(iv) The assumption in (iii) accepts the linearity and additivity of the model. If there are doubts about these assumptions, it may be better to take not a random-chosen donor but instead one close to the recipient in terms of his x-values. 'Close' may be defined by a distance function as in (g) above. In the limit, if a donor with the same set of x-values is found, this procedure reduces to assigning that donor's y-value to the recipient.

An important special case of the regression model with one control variable occurs when B_o is set equal to zero. In this case the model reduces to the ratio model $y_k = B_1 x_{1k} + e_k$. The ratio model may often be appropriate for panel surveys, with x representing the same variable as y measured on the previous wave of the panel. Ford _et_ _al_. (1980) investigate the use of a ratio model for imputation.

Through the careful development of an explicit model, regression imputation has the potential to produce imputed values closer to the true but missing values than the rough-and-ready, assumption-free, imputation class approaches. The construction and assessment of a good regression model is, however, a time-consuming operation, and it seems unrealistic to consider its application for all the items with missing values in a survey. If it is used at all, it seems best to reserve it for a few major variables for which effective linear models with high R^2's can be developed. In using regression imputation, attention also needs to be given to problems of estimating several missing items on the same record. This point is discussed further in Section 4.4.

Applications of regression and categorical data models for imputation are described by Schieber (1978), Herzog and Lancaster (1980), and Herzog (1980).

4.3 Two Simulation Studies

Two small-scale simulation studies have been carried out using data collected in the 1978 ISDP Research Panel to illustrate and compare some of the imputation procedures described in the previous section. One study compares methods of imputing for missing data on the variable hourly rate of pay for those jobs paid at an hourly rate, as reported in the July 1978 questionnaire. The other concerns the variable quarterly earnings, as derived from income records in the April 1978 questionnaire. Both studies are confined to the area probability sample of the 1978 ISDP Research Panel.

The approach used in both studies consisted of first constructing data sets with missing values by taking all the cases with valid data for the variable concerned and deleting some of the recorded values, and then applying a selection of imputation procedures to determine how well they reconstructed the deleted values. The data sets were constructed in a way that attempted to mirror the actual patterns of missing data observed in the full sample. Subgroups of the sample with differing rates of missing data were identified, and values were deleted from cases with valid data at the corresponding rates within these subgroups. It was also necessary to remove some cases with valid data entirely in order to produce data sets with the same distributions over the subgroups as that of the full sample. Within subgroups, the cases whose values were deleted were selected at random. As a protection against the selection of an odd sample of deleted cases, the process was replicated ten times producing ten simulation data sets for each study. Many of the measures evaluating the quality of the imputation schemes are averaged across the ten data sets. The choice of subgroups with differing rates of missing data for the two variables were based on the results of nonresponse analyses of the 1978 ISDP Research Panel reported by Heeringa (1980).

Once the data sets were constructed, a variety of imputation procedures were applied to fill in the deleted values. The control variables used in these procedures were selected on the basis of Hendrix's (1980) analyses of effective predictors for the two variables.[2] Imputation procedures with stochastic components were implemented ten times. Many of the evaluation measures reported below are averaged over these ten replications as well as over the ten simulation data sets.

While these simulation studies provide useful test-beds for investigating the various imputation procedures, their limitations should be recognized. In constructing the data sets attempts were made to reflect the actual patterns of missing data by deleting values at differing rates in various subgroups. However, ultimately within subgroups the cases for deletion were chosen at random, a feature which departs from reality to some, unknown, extent. The studies reported below

---

[2]For reasons of confidentiality geographical data were deleted by the Department of Health and Human Services from our data set. For this reason geographical variables were not available for use as imputation control variables.

directly examine how the imputation procedures perform in dealing with the missing data model employed, and only indirectly suggest how they might perform in actual practice. In particular, note that if the subgroups from which the values were deleted at differing rates were themselves used as imputation classes with the simulated data set, the estimates of means and totals of the variables would automatically be unbiased estimates of the corresponding population parameters. (Through the method of construction, the data are missing at random within these classes.) There can be no guarantee, however, that this also applies for the actual sample.

4.3.1 Imputation of Hourly Rate of Pay

The dependent variable for this study is hourly rate of pay (SC 1335-47-59) on the July questionnaire of the 1978 ISDP Research Panel. The analyses are conducted at the job level, so that, if a respondent has more than one hourly paid job, each job is taken individually (however, few employees in fact have more than one job). Only respondents in the area probability sample of the Research Panel who report having one or more hourly paid jobs are included in the analyses.[3]

The two characteristics most highly associated with the item nonresponse rate for hourly rate of pay are the person's sex and whether the interview was conducted in person or with a proxy informant. Table 1 gives the item nonresponse rates for the four cells comprising the cross-classification of these two characteristics.

The study's ten data sets were constructed in two steps from the 1128 cases with reported hourly rates of pay. First 92 cases were removed, randomly within cells, so that the distribution of the remaining cases conformed to that of the second column of Table 1. Next the missing data rates in the last column of Table 1 were applied to delete values at random within each of the four subgroups. This process was replicated ten times, producing the ten simulation data sets. The sex and

---

[3]There are some slight discrepancies between the results reported by Heeringa (1980) and those reported here, resulting from a minor variation in definition of cases covered, compounded with errors in the data set.

Table 1: Item nonresponse rates for hourly rate of pay by sex and interview status: Area Probability Sample, July 1978.

	Total hourly paid jobs		No. of jobs with hourly rate of pay reported	Percent of missing data
	No.	%		
Male, self report	339	27.0	327	3.5%
Male, proxy report	345	27.5	285	17.4%
Female, self report	432	34.4	400	7.4%
Female, proxy report	139	11.1	116	16.5%
Total[1]	1255	100.0	1128	10.1%

[1]There were also 9 cases with missing interview status, 8 of which had valid data on hourly rate of pay. These cases were excluded from the simulation study reported here.

Table 2: Distribution of simulated data sets by sex and interview status for hourly rate of pay imputation study

	Total hourly paid jobs		Hourly rate retained		Hourly rate deleted		Percent of deleted cases
	No.	%	No.	Mean rate ($)	No.	Mean rate ($)	
Male, self	280	27.0	270	5.23	10	5.49	3.6%
Male, proxy	285	27.5	235	5.05	50	5.31	17.5%
Female, self	356	34.4	330	3.52	26	3.60	7.3%
Female, proxy	115	11.1	96	3.39	19	3.35	16.5%
Total	1036	100.0	931	4.39	105	4.55	10.1%

interview status distribution of each of the data sets so constructed is given in Table 2. The table also gives hourly rates of pay averaged across the ten simulation data sets.

Of the 1036 cases in each constructed data set, 105 have values of the hourly rate of pay deleted. Several alternative imputation procedures have been applied to assign values for these deleted values. The imputation control variables employed in these procedures were obtained from a SEARCH analysis conducted by Hendrix (1980, Section 3.1) to identify good predictors of hourly rate of pay from the April wave data. SEARCH, the successor to the well-known Automatic Interaction Detector (AID) program, determines the most effective predictors of a variable through a series of binary splits of the sample, each split partitioning the sample into two parts on some predictor variable in such a way that the maximum proportion of variance is explained (for further details see Sonquist _et al_., 1974). The results of the analysis may be conveniently presented in a tree diagram. The tree diagram from Hendrix's analysis is reproduced in Figure 1.

The following imputation procedures were implemented:

1. Grand mean imputation (GM) assigning the respondent mean for all the missing values.

2. Class mean-value imputation using 8 imputation classes (CM8) defined by the cross-tabulation of the first three variables appearing in the SEARCH analysis, namely union membership (member v. not a member and membership not known), occupation (professional, technical and crafts v. others) and industry (agriculture, forestry, fisheries, mining and construction, transportation, communications and other public utilities v. others, i.e. Industry codes 017-077 and 407-479 v. others). The distribution of cases with retained and deleted values across these eight cells for the first simulation data set is given in Table 3.

3. Class mean-value imputation using 10 imputation classes (CM10) defined by the SEARCH analysis, that is the groups with '*'s in Figure 1, groups numbered 9 to 19, except for group 12. The distributions of cases with retained and deleted values across these ten cells are given in Table 4.

FIGURE 1

SEARCH ANALYSIS OF HOURLY RATE OF PAY [a]

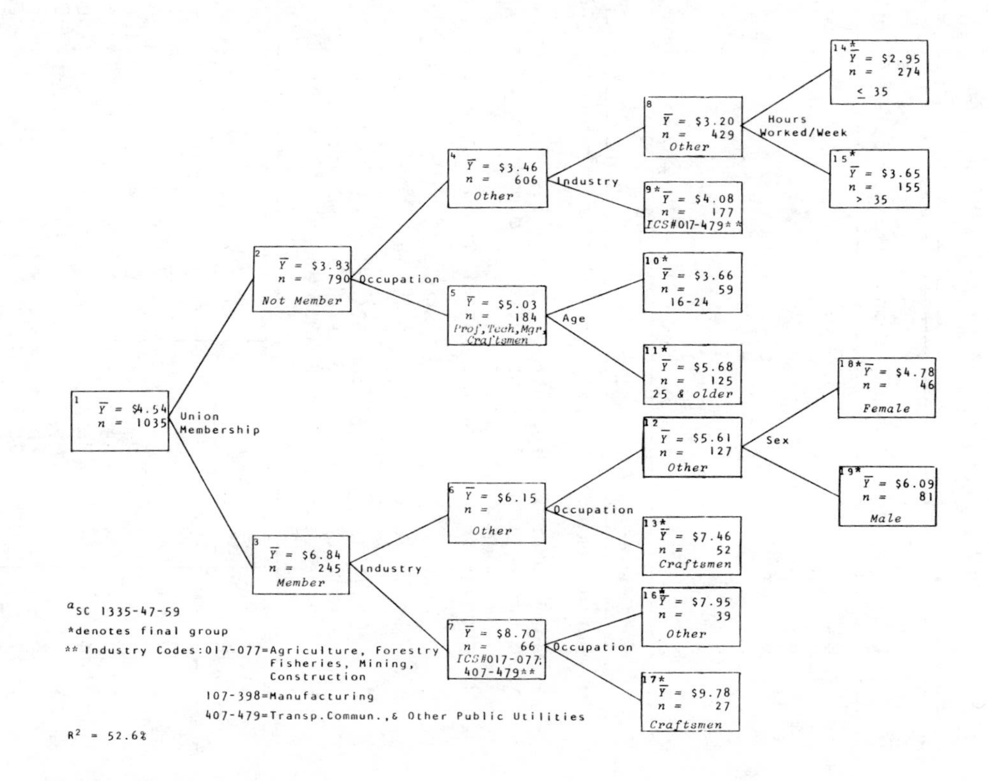

[a] SC 1335-47-59

*denotes final group

** Industry Codes: 017-077=Agriculture, Forestry Fisheries, Mining, Construction
107-398=Manufacturing
407-479=Transp.Commun., & Other Public Utilities

R^2 = 52.6%

Table 3: Distribution of cases in the first simulation data set with retained and with deleted values by union membership, industry and occupation

Union membership[1]	Industry[2]	Occupation[3]	Values retained	Values deleted	Total
M	A	P	24	6	30
M	A	O	22	5	27
M	O	P	46	6	52
M	O	O	97	16	113
N	A	P	27	4	31
N	A	O	55	4	59
N	O	P	117	14	131
N	O	O	543	50	593
		Total	931	105	1036

[1]M: union member; N: not a member or membership not known.

[2]A: agriculture, forestry, fisheries, mining and construction, transportation, communications and other public utilities; O: other.

[3]P: professional, technical, crafts; O: others.

Table 4: Distribution of cases in the first simulation data set with retained and with deleted values for the ten SEARCH analysis classes

Class[1]	Values retained	Values deleted	Total
9	157	17	174
10	50	6	56
11	94	12	106
13	38	4	42
14	226	24	250
15	215	13	228
16	26	5	31
17	20	6	26
18	38	6	44
19	67	12	79
Total	931	105	1036

[1]See Figure 1 for definitions of classes.

4. Random imputation within the 8 imputation classes (RC8) defined in procedure 2 above. The donors were chosen within classes by SRS.

5. Random imputation within the 10 imputation classes (RC10) defined by the SEARCH analysis as described for procedure 3 above. The donors were chosen within classes by SRS.

6. Multiple regression imputation using the predicted values from the regression equation (MI). The following independent variables were used in the regression: union membership as one dummy variable (member v. not a member or membership not known); occupation as 6 categories or 5 dummy variables (professional, technical, managers and administrators; sales and clerical; craftsmen; operatives; laborers and farm managers; occupation not known); industry as 6 categories or 5 dummy variables (agriculture, forestry, fisheries, mining, construction, transport, communications and other public utilities; manufacturing; wholesale and retail trades; finance, insurance, real estate, public administration; professional and related services; and industry not known); age as a grouped continuous variable (in four groups: 16-24; 25-44; 45-64; 65 and over); sex as a dummy variable; and hours worked per week as a continuous variable. The regression analysis of hourly rate of pay on these independent variables applied to all 931 cases in the first simulation data set with hourly rate of pay retained had an R^2 of 0.53.

7. Multiple regression imputation using the predicted value from the regression plus a random residual chosen from a normal distribution (MN). The procedure employs the same regression analysis as procedure 6. The random residuals are chosen from $N(0,s_e^2)$, where s_e^2 is the estimated residual variance for the respondents. The residual standard deviation for the first simulation data set was $s_e = 1.501$.

8. Multiple regression imputation using the predicted value from the regression plus a randomly chosen respondent residual (MR). This procedure employs the same regression analysis as that described in procedure 6. The values imputed for the deleted cases are the predicted values plus

a residual chosen at random by SRS from the respondents' residuals.

Three measures for evaluating the effectiveness of these eight imputation procedures in recapturing the deleted values are presented in Table 5. They are:

(a) The mean deviation defined as $\Sigma(\hat{y}_{mi} - y_{mi})/m$, where $\hat{y}_{mi}$ is the imputed value and y_{mi} is the actual, but deleted, value for case i (i = 1,2,...m).

(b) The mean absolute deviation defined as $\Sigma|\hat{y}_{mi} - y_{mi}|/m$.

(c) The root mean square deviation defined as $\{\Sigma(\hat{y}_{mi} - y_{mi})^2/m\}^{1/2}$.

These measures are averaged across the ten simulation data sets. The procedures RC8, RC10, MN and MR involve a stochastic component. For each of these procedures, the measures are also averaged over ten replications of the imputations.

Table 5: Measures of the effectiveness of the eight imputation procedures for imputing for deleted hourly rates of pay

Imputation scheme	Mean deviation (¢)	Mean absolute deviation (¢)	Root mean square deviation (¢)
1. GM	-15.8	190.4	252.7
2. CM8	-2.7	126.1	173.0
3. CM10	-1.9	120.3	167.3
4. RC8	-5.2	172.9	244.6
5. RC10	-0.3	165.8	236.2
6. MI	-4.9	119.6	170.0
7. MN	-4.7	175.3	228.1
8. MR	-5.6	167.8	227.5

The mean deviation measures the bias in the imputed values. Since the cases with missing values are only part of the total sample, the resulting bias in the estimation of the population mean is only a fraction $\bar{M}$ (the nonresponse rate) of the mean deviation. As seen in Table 5, all the mean deviations are negative, indicating that on average the imputed values underestimated the actual (deleted) values. The average underestimation for the grand mean (GM) procedure is almost 16 cents, and this leads to the same bias in the estimate of the population mean as would the procedure of simply computing the respondent mean ignoring the missing values. The average underestimation for the other procedures is less than 6 cents, so that the use of the control variables seems to have been effective in reducing the absolute magnitude of the imputation bias, as compared with the GM procedure which does not make use of any control variables.

The mean absolute deviations and the root mean square deviations measure the "closeness" with which the deleted values are reconstructed. Necessarily, the GM procedure fares worst in terms of these indices, having the largest measures on both of them. Table 5 shows that the three procedures using predicted values without residuals (CM8, CM10, and MI), yield the best measures, all around 120 for the mean absolute deviation and 170 for the root mean square deviation. As expected, the CM10 procedure does slightly better than the CM8 procedure, indicating the gains from increasing the extent of use of the control variables.

The four procedures that incorporate another stage of randomization through the addition of a residual term in some way (RC8, RC10, MN, and MR) yield mean absolute and root mean square deviations that are approaching 40 per cent larger than those for their deterministic counterparts (CM8, CM10 and MI). As the following argument demonstrates, this result is very close to what might be reasonably predicted.[4] Consider the mean square deviation of the RC procedure given by

$$\mathrm{MSD(RC)} = \Sigma\Sigma(\hat{y}_{mhi} - y_{mhi})^2/m.$$

Then

$$\mathrm{mMSD(RC)} = \Sigma\Sigma(\hat{y}_{mhi} - \bar{y}_{rh})^2 + \Sigma\Sigma(y_{mhi} - \bar{y}_{rh})^2 + \Sigma\Sigma(\hat{y}_{mhi} - \bar{y}_{rh})(y_{mhi} - \bar{y}_{rh})$$

[4] I am grateful to Brian Greenberg and James Fagan for drawing my attention to this point.

Taking the conditional expectation over the random imputations within each imputation class, $E_2(\hat{y}_{mhi}) = \bar{y}_{rh}$ so that for large r_h,

$$mE_2[MSD(RC)] \simeq \Sigma m_h s_{rh}^2 + mMSD(CM)$$

This equation shows that $E_2[MSD(RC)] \geq MSD(CM)$. Under the assumption that values are missing at random within imputation classes, $MSD(CM) \simeq \Sigma m_h s_{rh}^2$, so that $E_2[MSD(RC)] \simeq 2MSD(CM)$. Thus, under the missing at random within classes assumption, the root mean square deviation of the RC procedure can be expected to be about $\sqrt{2} = 1.41$ times as large as the CM imputation procedure that employs the same set of imputation classes. Under appropriate assumptions, the corresponding factor of $\sqrt{2}$ can also be derived for the stochastic and deterministic regression imputation procedures.

The reason for accepting the less accurate estimates from the imputation procedures with the residual terms is that these residuals maintain the distributional properties of the sample. This case is demonstrated in Table 6 which summarizes the distributions of the imputed values for the eight procedures, and provides the distribution of the true, but deleted, values (TV) for comparison. The table relates only to the first simulation data set, and only to one replication of the imputation schemes incorporating a stochastic component.

The first three columns of Table 6 clearly illustrate the distortion to distributions created by mean-value imputation. The GM procedure allocates the respondent sample mean to all the missing cases, thus concentrating all the distribution of missing values at the one value. The number of different imputed values that can occur with the CM procedures is restricted to the number of imputation classes, and the number of cases with any one of these values is then the number of cases with missing data in the given class. As Table 3 shows, the last class in procedure CM8 is extremely large, containing over half the sample. Fifty of the cases in this class have missing data, and they are all assigned a value of \$3.20: these cases thus account for the excessive number of cases in the \$3-\$4 interval in the CM8 column of Table 6. The distribution for the regression imputation procedure using

Table 6: Distributions of the true values (TV) and of the imputed values from the eight imputation procedures for hourly rate of pay of the 105 cases with deleted values in the first simulation data set.

Hourly Rate of Pay ($)	TV	Imputation Procedure'							
		GM	CM8	CM10	RC8	RC10	MI	MN	MR
$2 or less	3	-	-	-	1	5	2	12	10
$2-	22	-	-	24	29	24	13	17	15
$3-	28	-	50	19	24	25	33	14	21
$4-	11	105	22	23	10	10	14	15	16
$5-	11	-	16	24	15	8	13	12	13
$6-	10	-	-	-	7	8	16	15	13
$7-	10	-	11	9	13	16	14	16	13
$9-	3	-	6	6	2	6	-	4	3
$11-	4	-	-	-	4	3	-	-	1
$13 and over	3	-	-	-	-	-	-	-	-
Total	105	105	105	105	105	105	105	105	105

'For each imputation procedure the distribution is obtained from a single replication.

predicted values, MI, also shows signs of curtailment of spread, but this is less evident than in the case of the CM procedures; the lesser degree of bunching in this distribution is accounted for partly by the greater number of classes used (via dummy variables) and partly by the use of continuous variables in the regression. The MN and MR procedures, which add a residual term to the predicted value from the regression, produce substantial excesses in the numbers of very small hourly rates of pay. This outcome arises because of the inapplicability of the regression assumptions adopted. Both procedures assume that the residuals are homoscedastic, and the failure of this assumption may well account for the excess. Failure of the assumption of normality of residuals may also be a contributory factor in the case of the MN procedure. The two RC procedures generate the best distributional properties among the procedures considered here.

4.3.2 Imputation of Quarterly Earnings

In the 1978 ISDP Research Panel, respondents were asked to provide information on their earnings from records where possible. If they were paid the same amount each payday, they were asked to report the amount of the paycheck and the frequency of pay. If they were paid different amounts each payday, the amounts were collected for each of the paychecks received in the preceding three months. From these items of information an aggregate quantity, quarterly earnings, can be constructed. This quantity for the period January to March 1978, as reported in the April 1978 wave of the 1978 Research Panel, is the dependent variable for the second study. The data are analyzed at the job level, and the analyses relate to respondents in the area probability sample only.

The designers of the 1978 Research Panel questionnaires anticipated that it would often not prove possible to collect earnings data from records. They therefore included alternative questions in the questionnaire to serve as the basis for estimating earnings. In particular, data were collected on hourly rates of pay, usual numbers of hours worked per week, and numbers of weeks without pay in the last quarter. These data, which can then be used to estimate quarterly wages, are employed below in a form of deductive imputation.

The methods adopted for the study of imputation procedures for quarterly earnings are the same as those used for hourly rates of pay described in the last section. The two characteristics most highly associated with the missing data rate on quarterly earnings reported from records were interview status and the household's monthly income for March (SC 0014).[5] The missing data rates for the four cells defined by interview status and by whether the March household income was under $900 or not are given in Table 7. The table shows that, in more than half the cases, the respondent did not provide earnings data from records.

The ten simulated data sets used as the basis of the study of imputation procedures were constructed in the manner described in the previous section. The distribution of each of these data sets over the four cells contained in Table 7 is

---

[5]The household monthly income question was asked early in the interview, and was used for screening purposes. It was not intended to provide an accurate measure of monthly income, and hence cannot be used for calculating quarterly earnings.

presented in Table 8. The mean quarterly earnings presented in Table 8 are averages over the ten simulation data sets.

The choice of the imputation control variables to be employed were based on a SEARCH analysis of quarterly earnings carried out by Hendrix (1980, Section 3.2). The tree diagram summarizing his results is presented in Figure 2.

Ten different imputation procedures were applied to impute for the deleted values:

1. Grand mean imputation (GM) assigning the respondent mean for all the missing values.
2. Class mean-value imputation using 8 imputation classes (CM8). Seven of these classes were obtained from the crosstabulation of the first three variables in the SEARCH analysis - sex, household's March income (under $900, $900 or more) and work status (full- or part-time). The cell of this crosstabulation containing men part-time workers with household income under $900 contained only two cases with earnings retained but 11 cases with earnings deleted: this cell was therefore collapsed with that for women part-time workers with household income under $900. The eighth cell comprised a small number of cases with one or more of the classification variables missing. The distribution of cases with earnings retained and those with earnings deleted across the eight cells for the first simulation data set is shown in Table 9.
3. Class mean-value imputation using 12 imputation classes (CM12) defined by the SEARCH analysis, that is the groups with '*'s in Figure 2, groups numbered 8, 11, 12, 14, and 16 to 23. The distributions of cases with retained and deleted values across these 12 classes are given in Table 10.
4. Random imputation within the 8 imputation classes defined in procedure 2 above (RC8). The donors were chosen within classes by the SRS scheme described earlier in the report. As Table 9 demonstrates, in many classes there were more recipients than potential donors. In these classes each case with a retained value served as a donor once, being allocated to recipients at random, and a SRS of cases with

Table 7: Missing data rates for quarterly earnings reported from records by interview status and household's March income: Area Probability Sample, April 1978[1]

Household's March Income	Interview Status	Total jobs No.	Total jobs %	No. of jobs with earnings reported	Percent of missing data
Under $900	Self	404	21.7	188	53.5%
Under $900	Proxy	176	9.5	38	78.4%
$900 or more	Self	771	41.5	424	45.0%
$900 or more	Proxy	507	27.3	191	62.3%
	Total	1858	100.0	841	54.7%

[1]67 cases with either missing interview status or missing household March income are excluded from this table, and from the following simulation study.

Table 8: Distribution of simulated data sets by household's March income and interview status for quarterly earnings imputation study

Household's March income	Inter-view status	Total jobs		Quarterly earnings retained		Quarterly earnings deleted		Percent of deleted cases
		No.	%	No.	Mean ($)	No.	Mean ($)	
Under $900	Self	87	21.7	40	1384	47	1421	54.0%
Under $900	Proxy	38	9.5	8	2351	30	1995	78.9%
$900 or more	Self	166	41.5	91	3135	75	3380	45.2%
$900 or more	Proxy	109	27.3	41	3152	68	3116	62.4%
	Total	400	100.0	180	2715	220	2691	55.0%

FIGURE 2

SEARCH ANALYSIS OF WAGE AND SALARY INCOME [a]

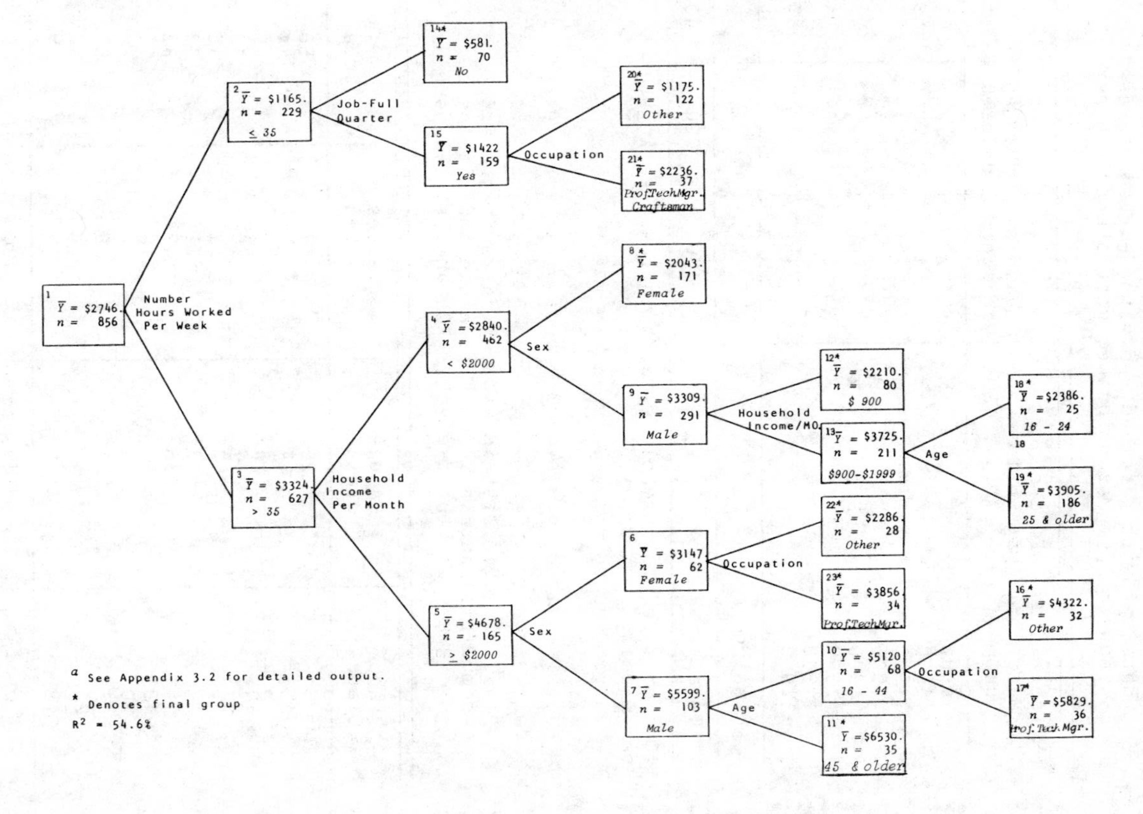

[a] See Appendix 3.2 for detailed output.

* Denotes final group

R^2 = 54.6%

Table 9: Distribution of cases in the first simulation data set with retained and with deleted values by household's March income, work status and sex

Household's March income	Full-or part-time F/P	Sex	Values retained	Values deleted	Total
< $900	F	M	23	27	50
< $900	F	F	11	22	33
< $900	P	M	2 }	11 }	13 }
< $900	P	F	12 }	17 }	29 }
≥ $900	F	M	65	79	144
≥ $900	F	F	35	33	68
≥ $900	P	M	11	12	23
≥ $900	P	F	19	17	36
Missing			2	2	4
Total			180	220	400

[1]One or more of household's March income, work status and sex missing.

Table 10: Distribution of cases in the first simulation data set with retained and with deleted values for the twelve SEARCH analysis classes

Class[1]	Values retained	Values deleted	Total
8	23	27	50
11	7	7	14
12	8	7	15
14	7	8	15
16	7	8	15
17	13	24	37
18	36	49	85
19	34	40	74
20	9	9	18
21	24	26	50
22	8	10	18
23	4	5	9
Total	180	220	400

[1]See Figure 2 for definitions of classes.

retained values was selected to donate values to the remaining recipients (in the second row of the table - the <$900, F,F class - each case with a retained value was in fact used twice as a donor).

5. Random imputation within each of the 12 imputation classes defined by the SEARCH analysis as described in procedure 3 above (RC12). The donors were chosen by the SRS scheme described for procedure 4.

6. Multiple regression imputation using the predicted values from the regression equation (MI). The following independent variables were used in the regression: age as a continuous variable; number of hours worked per week; whether job was held for the full quarter or not, as a single dummy variable; occupation as 5 categories or 4 dummy variables (professional, technical, managers and administrators; sales and clerical; craftsmen; operatives; laborers and farm managers and occupation not known); household March income as 3 categories or 2 dummy variables (under $900; $900-$1999; $2000 or more); and sex as a dummy variable. This regression analysis, applied to 176 of the 180 cases with quarterly earnings retained in the first simulation data set (the other four cases were dropped because they had some missing data on the independent variables), yielded an R^2 of 0.58.

7. Multiple regression imputation using the predicted value from the regression plus a random residual from a normal distribution (MN). The procedure employs the same regression analysis as procedure 6. The random residuals are chosen from $N(0,s_e^2)$, where s_e^2 is the estimated residual variance for the respondents. The residual standard deviation for the first simulation data set was $s_e = 1,367$.

8. Multiple regression imputation using the predicted value plus a randomly chosen respondent residual (MR). This procedure employs the same regression analysis as procedure 6. The values imputed for the deleted cases are the predicted values plus residuals chosen at random by SRS from the respondents' residuals.

9. Mixed deductive and random imputation using 8 imputation classes (DI8). For about 60 per cent of cases with earnings deleted, information was available to estimate their quarterly earnings by multiplying their hourly rate of pay by the number of hours usually worked per week by the number of weeks worked in the quarter; for these cases, the missing values were imputed from this formula. The imputed values for the remaining 40 per cent of cases were obtained by random imputation with SRS within the eight imputation classes defined in procedure 2.
10. Mixed deductive and random imputation using the 12 imputation classes defined by the SEARCH analysis - see procedure 3 (DI12). This procedure is the same as the previous one except that the random imputations for the approximately 40 per cent of cases were carried out within these twelve classes rather than the eight classes used in procedure 9.

The mean deviations, mean absolute deviations, and root mean square deviations measuring the effectiveness of these ten procedures in imputing for the 220 deleted values, are presented in Table 11. These measures are averaged across the ten simulation data sets and, in the case of the imputation procedures involving a stochastic component, also over ten replications of the procedure with each data set.

Table 11: Measures of the effectiveness of the ten imputation procedures for imputing for quarterly earnings

Imputation scheme	Mean deviation ($)	Mean absolute deviation ($)	Root mean square deviation ($)
1. GM	24.3	1571	2028
2. CM8	-29.2	1103	1609
3. CM12	-18.7	1033	1480
4. RC8	-28.5	1538	2191
5. RC12	-21.3	1379	1964
6. MI	-3.2	1046	1440
7. MN	-7.1	1520	1965
8. MR	-1.7	1491	1972
9. DI8	-149.5	1052	1737
10. DI12	-97.4	955	1564

The mean deviations in Table 11 vary between the various imputation procedures. From theoretical considerations, the mean deviations of the three regression procedures (MI, MN and MR) should be similar and the class mean-value imputations (CM8 and CM10) should have similar mean deviations to their stochastic counterparts (RC8 and RC10 respectively). The results in Table 11 conform well to these expectations. The regression procedures produce almost unbiased estimates, while the class mean-value procedures have negative biases of around $20 to $30. The grand mean procedure, however, yields a positive bias. All the mean deviations are relatively small, around 1% or less, except for the two mixed deductive and random class imputation procedures (DI8 and DI12). The relatively larger biases of these latter procedures will be discussed later.

Comparison of the mean absolute deviations and root mean squared deviations between the deterministic imputation procedures (CM8, CM12 and MI) and their stochastic counterparts (RC8, RC12 and MN and MR, respectively) shows that these evaluation measures are on average over one third larger for the stochastic procedures (this is somewhat less than the 41% larger predicted by the appropriate missing at random assumption as discussed in the Section 4.3.1). Of these procedures, the CM10 and MI procedures fare best, with the CM8 procedure being a little less good. As a result of its lack of use of control variables, the deterministic GM procedure has mean absolute and root mean squared deviations of about the same size as the stochastic procedures. Although the two mixed deductive and random imputation procedures, DI8 and DI12, incorporate a stochastic component in imputing for about 40 per cent of cases, they have mean absolute and root mean squared deviations of about the same size as the corresponding class mean-value procedures CM8 and CM12. This finding suggests that the use of the deductive imputation with the other 60 per cent of cases must yield imputed values close to the actual values, a suggestion substantiated in Table 12 where the two sets of cases are taken separately.

The deductive imputation in the DI8 and DI12 procedures produces a mean absolute deviation of about 490 for the cases

Table 12: Mean deviations and mean absolute deviations for the ten imputation procedures for imputing quarterly earnings separately (a) for the cases for which the deductive imputation based on hourly rate of pay, usual number of hours worked per week and number of weeks worked in the quarter can be made and (b) for the cases for which this imputation cannot be made

Imputation scheme	Mean deviation		Mean absolute deviation	
	(a) for deductive cases ($)	(b) for nondeductive cases ($)	(a) for deductive cases ($)	(b) for nondeductive cases ($)
1. GM	563	-692	1423	1674
2. CM8	181	-343	964	1294
3. CM12	181	-253	934	1127
4. RC8	187	-361	1342	1773
5. RC12	196	-267	1254	1511
6. MI	65	-85	870	1275
7. MN	68	-67	1394	1689
8. MR	56	-105	1348	1682
9. DI8	94	-361	494	1773
10. DI12	89	-267	489	1511

with which it can be used. This value is considerably lower than the mean absolute deviations for any of the other procedures for these cases. In these terms, the next best procedure is the regression prediction (MI), and its mean absolute deviation is about 77 per cent larger than that for the deductive imputation. The use of deductive imputation, or at least the inclusion of the deduced values (or some or all of their components) in the imputation procedure employed, makes a valuable contribution in imputing for quarterly earnings.

The separation of deductive and non-deductive cases in Table 12 also sheds further light on the mean deviations for the various procedures. A striking finding is that the mean deviations are positive for the deductive cases and negative for the non-deductive cases for all ten procedures. These fairly sizeable positive and negative measures tend to cancel out in the aggregate, creating the small mean deviations observed in Table 11. For the non-deductive cases, the use of the DI8 and DI12 procedures means employing the RC8 and RC12 procedures respectively, so that the mean deviations of the DI8 and RC8 and of the DI12 and RC12 procedures are the same for these cases (and similar also to the mean deviations for the CM8 and CM12 procedures). The deductive imputation procedures, however, yield lower mean deviations for the deductive cases than their RC (and CM) counterparts. As a result, the cancellation of positive and negative mean deviations between deductive and non-deductive cases that applies in the aggregate with the RC and CM procedures does not occur to the same extent with the DI procedures. This explains the larger mean deviations of the DI procedures seen in Table 11.

The deductive imputation, which consists of multiplying hourly rate of pay by number of hours usually worked per week by number of weeks worked in the quarter, applies only to wage earners for whom these data are available. Thus all the deductive cases are wage earning jobs and, on examination, the non-deductive cases are found to be mostly salaried jobs. Thus it appears that all the imputation procedures implemented in this study tend to overestimate wages and underestimate salaries. Such biases would not occur if the data were missing at random within the imputation classes employed in the CM and RC procedures. The missing data were in fact generated at random within the four subgroups given in Table 7, subgroups created by cross-classifying household's March income (under $900, $900 or more) and interview status (proxy v. self report). Since the imputation classes used in the CM8 and RC8 procedures included the household's March income in exactly the same form, and this income was also partially included in defining the SEARCH imputation classes used in the CM12 and RC12 procedures, it seems that the biases have mainly occurred because of the failure of the imputation classes to control on the interview status variable.

The distributions of the imputed values for the ten procedures are compared with the distribution of the actual, but deleted, values in Table 13. As with the first study, the GM procedure concentrates all the imputed values at one point and the CM8 and CM12 mean-value procedures create distributions with spikes at the imputation class means. The procedure using the predicted values from the regression (MI) produces a bunching of imputed values in the middle of the distribution. The distributions for the RC8 and RC12 procedures fit the distribution of true values reasonably well, as also do the DI8 and DI12 procedures (but their fits seem to be not quite as good).

Table 13: Distributions of the true values (TV) and of the imputed values from the ten imputation procedures for quarterly earnings for the 220 cases with deleted values in the first simulation data set

Quarterly earnings	TV	Imputation procedure'									
		GM	CM8	CM12	RC8	RC12	MI	MN	MR	DI8	DI12
< $0	-	-	-	-	-	-	2	19	23	-	-
$0-	50	-	45	50	52	47	9	26	20	55	52
$1000-	48	-	34	-	42	42	30	39	34	36	33
$2000-	48	220	60	88	40	45	39	33	52	50	54
$3000-	25	-	2	59	30	34	41	42	39	31	38
$4000-	20	-	79	8	28	26	56	29	25	22	17
$5000-	13	-	-	-	12	12	27	21	15	11	11
$6000-	9	-	-	15	9	7	12	7	5	11	8
$7000-	3	-	-	-	4	3	4	4	5	2	4
$8000+	4	-	-	-	3	4	-	-	2	2	3
Total	220	220	220	220	220	220	220	220	220	220	220

'For each imputation procedure the distribution is obtained from a single replication.

The two regression procedures that employ predicted values plus residuals, MN and MR, encounter the problem of producing a sizeable number of negative imputed values. One solution for handling these impossible values would be to treat them as zero, a solution which would clearly lower the mean absolute and root mean square deviations. It would increase the mean deviations, turning negative deviations into positive ones. Treating the negative values as zero also improves the fit in the $0-$999 cell, raising the numbers in that cell to 45 for the MN procedure and to 43 for the MR procedure. An alternative solution would be to take a transformation of the earnings variable initially, to impute the values of the transformed variable, and then to transform the imputed values back into earnings amounts; for instance, a logarithmic transformation would ensure non-negative imputed earnings amounts. The transformation approach may also often improve the fit of imputed values to highly-skewed distributions of actual values. Serious consideration should be given to the use of transformed variables in imputation when a regression model is contemplated.

The main purpose of presenting the results of the two studies in this section is to illustrate the effects of a variety of imputation procedures on some univariate statistics (Santos, 1981a, uses the same simulation studies to investigate the effects on multivariate statistics). The particular applications chosen were not developed in great detail to be the optimum versions of the specific procedures; in consequence, the results should be interpreted with caution, bearing this point in mind. Nevertheless, there are some general conclusions that can be tentatively drawn.

First, the studies illustrate the benefit and cost of using random rather than mean-value imputation within imputation classes. The benefit can be clearly seen from the comparisons of the distributions for the two types of procedures, with the random imputed values having distributions reasonably close to those of the true values, while the mean-value imputations have distorted, spiked, distributions. The cost is seen by comparing the mean absolute deviations or the root mean square deviations for the two types of procedure, where the measures for random imputations are about a third or more higher than those for the mean-value imputations. While the amount of increase in these measures depends on the homogeneity of the imputation classes employed, and on the dependent variable under study, it seems likely that increases of at least this magnitude could often occur in practice.

Secondly, the results indicate the security derived from using a random imputation procedure rather than a regression model of the type employed. The failure of the data to conform to the assumptions made about the distribution of residuals in the stochastic regression models (MN and MR) created some distortions in the distributions of imputed values. The lack of need for specific distributional assumptions in the random imputation procedures enabled these distortions to be avoided. It should be made clear that the argument here is not against all uses of regression or other such models, but rather to point out the need to carefully construct and test out such models if they are to be employed. Unless sufficient resources are put into the development of the models to ensure that they do adequately reflect the distributional properties of the data, it is probably safer to apply an imputation procedure that assigns donors' values to recipients. This is not intended to suggest that such imputation procedures are assumption free, for they do depend on the assumption that data are missing at random within imputation classes. They therefore require an appropriate choice of imputation classes. However, in assigning donors' values to recipients, they automatically preserve the distribution of the donors' values (within chance variations), and hence avoid the need for assumptions specifying a more general distribution form.

Finally, the relative success of the deductive imputation procedure using hourly wage rates, usual numbers of hours worked per week and numbers of weeks worked for estimating quarterly wages brings out the wisdom of anticipating those survey variables that are likely to encounter high levels of missing data, and collecting substitute information which may be used in imputation - whether deductive or some other form of imputation is used.

4.4 Other Imputation Issues

Earlier sections of this report have sought to demonstrate the close relationship between weighting adjustments and imputation. It is now time to emphasize the difference between the two approaches, and hence to identify the added complexities of imputation.

Weighting class adjustments are appropriate when little important information is available for the nonrespondents. That information is incorporated in forming the weighting classes (or in raking ratio estimation), and any other information known about the nonrespondents is ignored. Weighting adjustments can be thought of as imputing complete sets of data from respondents to nonrespondents. Thus, for instance, in the equal weighting adjustment procedure within weighting class h (with r_h respondents) each nonrespondent's record can be thought of as being divided into r_h parts, each with a weight of $1/r_h$. Each part is then assigned the complete set of responses from one respondent. An important property of the resultant data set is that within weighting classes the data retain all the interrelationships between the survey variables that were found for the respondents, as is appropriate under the assumption that the missing data were missing at random within weighting classes (see Section 2.3.3). Moreover, providing the respondents' responses were consistent, satisfying any edit checks, the weighted data set is also bound to be consistent.

The position with imputation is made much more complicated by the fact that a great deal is known about the units with missing data, much more than can be incorporated in a matching of donors and recipients. Considering imputation for missing data on only a single item, it is necessary to distinguish between variables used in the imputation process (usually in forming imputation classes) and other variables. Variables used in the imputation process then need to be subdivided into variables for which full details are used and those for which only partial details are used (e.g. by categorizing a quantitative variable or combining categories of a qualitative variable). The interrelationships between the item with missing data and variables used in full detail in forming imputation classes have the desirable properties of a weighting adjustment procedure. However, the interrelationships between the item and variables not used in the imputation process may be distorted and the imputed values may fail edit checks involving the item and these variables. Similar damaging effects may also occur with variables used only partially in forming imputation classes.

A particular case of the problem of imputed values failing edit checks occurs when the item concerned is a quantitative one, and is related to a number of other quantitative variables. It is usually then impossible to obtain exact matches between donors and recipients in terms of these related variables, and approximate matches may result in the imputed values failing edit checks. An example will clarify this point. Suppose that the item with missing data is length of first marriage for persons known to have been married more than once, with the time from the date of the first marriage to the date of the second marriage being known. A matching of donors and recipients on the approximate time between first and second marriages may result in the lengths of first marriage imputed to some recipients exceeding the time interval between their two marriages. Thus, for instance, if a recipient with 22 years between marriages is matched with a donor with 25 years between marriages, the recipient may be allocated from the donor a length of first marriage of say, 23 years, an impossible length since it exceeds the interval between the recipient's two dates of marriage. A useful way to avoid this inconsistency is to impute not for the length of the first marriage but for this length as a proportion of the interval between the two marriages. In this way the imputed value is certain to satisfy the edit constraint. Inequalities of the type $A + B + C \leq D$ are often met with quantitative variables. In such cases a transformation of the type described may be appropriate (see, for instance, I.G. Sande, 1979b).

Another commonly encountered feature of quantitative survey variables is that they do not apply, or have a zero value, for many members of the sample. Thus, for instance, many people do not have investment and property incomes, and do not receive SSI payments. For such variables, it is often useful to consider imputation in two stages: first to impute whether the variable is zero or not; and then, if not zero, to impute the amount. Herzog (1980) uses this two-stage approach with regression imputation for the amount of Social Security benefits received. The "zero-spike" procedure of Ford *et al*. (1980) also employs this two-stage approach. Sometimes the first-stage answer is known from a filter question; in this case, if the amount is not zero, the imputation should be based on those with non-zero amounts.

An additional complexity with imputation is that every one of the survey variables is potentially subject to missing data. One difficulty this creates is that the variables used for defining the imputation classes for a given item might themselves sometimes be missing. A further issue arising in this case is that when two or more items are missing on a record it is preferable, *ceteris paribus*, to impute the values for them from the same donor; otherwise the covariance between the items will be attenuated and inconsistent values may be imputed. A way to carry out joint imputations is to employ the same imputation classes for all the items concerned and then to use the same donor for all the missing items for a given recipient. However, this procedure may operate against a choice of imputation classes that maximizes the proportion of variance explained in an individual item. Often a compromise solution is necessary, making joint imputations for a group of closely-related items, but treating different groups of items separately. One approach is a sequential procedure (Coder, 1978; Brooks and Bailar, 1978, Section IVB): first, fill in the "small holes" in basic items which are to be used as initial imputation classes; second, impute for a group of closely-related items using one set of imputation classes; third, impute for another group of variables using a different set of imputation classes (which may or may not be defined to include variables from the first group of variables); and so on.

The almost infinite range of possible patterns of missing data in any survey record makes it impracticable to construct individual procedures for each pattern. Instead, some small number of procedures is needed to cover all the possibilities. Thus, for instance, a single set of imputation classes may be defined for all items in one topic area and then imputations for any missing data for the items in that area are done simultaneously from a single donor. If a record has missing responses for all items in the topic area, this procedure reflects the interrelationships between items in the donor's record. However, if some responses are present and others are absent, the procedure attenuates the covariances between those imputed and those reported, and may possibly give rise to inconsistencies. One way to avoid these outcomes is to incorporate the reported values in the topic area in defining the imputation classes for the missing values. However, as already observed, this solution is usually impracticable because of the many different patterns of responses and nonresponses to the set of items. Another solution is to assign the donor's values for all items to a recipient, irrespective of whether the recipient's responses are totally or only partially missing. This solution is, of course, harmful for estimating univariate statistics for the various items, but is preferable for estimating the relationships between the items. If there are several items in the topic area, and only one or two responses are missing, this solution may be unattractive; if, however, only one or two responses are present, their deletion may create no serious loss of data.

When regression imputations with predicted values plus donor's residuals are used for several items, it may in the same way be advantageous to take residuals from the same donor for all of a recipient's imputed values. When such regression imputations are used for some items and some form of imputation class procedure is used for other items, there may again be benefits in using the same donor for a given recipient where possible. Procedures could easily be devised to achieve this objective.

The dissimilarities between weighting adjustments and imputation result in different considerations being involved in choosing imputation and weighting classes. As pointed out at the end of Chapter 3, the range of potential control variables from which weighting classes can be constructed is usually strictly limited, mainly to sample design variables and to variables whose distributions are available from an external source such as the Census. To be effective for a particular survey variable, weighting classes need to differ both in response rates and in the mean values of that variable. A typical survey collects data on many variables, each of which has a different set of relationships with the potential control variables, while on the other hand there is a single unit response/nonresponse variable that applies for all of the survey variables. For this reason the response variable plays a major role in determining weighting classes.

The imputation position differs in several respects. First, many more potential control variables are available. In addition to the control variables that may be used in forming weighting classes, all the other survey responses are available for use in forming imputation classes for a particular variable. This wealth of potential control variables highlights the need for an effective method for determining the combination of control variables to be used in defining imputation classes. Secondly, different sets of imputation classes can be considered for each survey variable (although, for the reasons already given, it may be preferable to use the same sets of classes for groups of variables). Thirdly, while the unit nonresponse rate is constant across all the survey variables, the item nonresponse rate changes from one variable to the next. And finally, as noted above, while weighting adjustments automatically retain relationships between items and maintain consistency, imputation does not do so. One consideration in forming imputation classes may therefore be the retention of certain relationships and the avoidance of certain inconsistencies.

As a consequence of these features of imputation, the prime consideration in forming imputation classes is the generation of good estimates of the missing values, estimates that are close to the actual values and that are consistent with other values on the record. There are two requirements for the achievement of this goal: one is that within each imputation class the donors' values should represent the recipients' values - i.e. the data missing at random assumption should hold - and the other is that within each class the donors' values should have little variance. In the ideal imputation classes, the values - both potential donors' and recipients' - have no variance. In this case, the imputation procedure is in effect a type of deductive imputation with the values being imputed without error.

In practice there is usually little information to guide the choice of imputation classes to satisfy the missing at random assumption, and therefore attention is mainly focussed on forming classes within which the potential donors are as homogeneous as possible with regard to the survey variable being considered. The general principle is then to form imputation classes that minimize the variance of the survey variable within classes, or equivalently that maximize the variance between classes. This is the principle behind the SEARCH technique - a successor to the Automatic Interaction Detector (AID) technique - used for determining imputation classes in the last section. In general the SEARCH technique seems a valuable tool for guiding the choice of imputation classes. However, since the technique capitalizes on chance patterns in the data, it should be applied with caution. The technique requires large samples, and splitting should not be continued to the point that the ultimate classes contain small numbers of cases. (The SEARCH technique can also be used for determining weighting classes. In this case, given the importance attached to the response variable, it may be used as the dependent variable in the analysis. See, for instance, Chapman, 1974).

The requirement that the same set of imputation classes is to be used as the basis for imputing for a group of variables will necessitate some compromise between the classes derived by SEARCH for each of the variables taken individually. If the variables in the group are closely related to each other, however, similar classes are likely to be obtained from the separate SEARCH analyses, and a compromise may not be hard to reach. Another possible approach is to employ a multivariate version of SEARCH, seeking to maximize the variance between classes with respect to a set of variables. Gillo and Shelly (1974) describe a multivariate version of SEARCH known as MAID - M (Multivariate Automatic Interaction Detection - Monitored).

A further consideration in forming imputation classes is that the sample should not be too finely divided into many small classes. The reasoning is the same as that applied to weighting classes, namely that the resultant instability in the compensations gives rise to a greater increase in variance than can be offset by the likely decrease in bias. With weighting, the argument may be framed in terms of variation in the weights, with large weights occurring by chance for some classes. The corresponding argument with imputation may be put in terms of variation in the extent of use of donors, with donors in some classes being used several times. Multiple use of some donors needs to be carefully controlled. Monitoring the usage of donors should be an important component of an imputation auditing process.

Finally, we should note in this chapter the close relationship between editing and imputation (I.G. Sande,

1979a). When edit checks identify inconsistencies, some item responses are deleted and replaced by imputed values. In cases when many interrelated edit constraints are specified, as for instance often occurs with industrial statistics, the choice of which values to delete when inconsistencies are found is a complicated one. A principle, such as minimizing the number of deletions, may be required (Greenberg, 1981; Fellegi and Holt, 1976). Another aspect of the relationship between editing and imputation is the need for the imputed values to satisfy the edit constraints. When many constraints are employed, the choice of imputed values to satisfy them may be severely limited. Hill (1978) discusses automatic edit and imputation in the 1976 Canadian Census with categorical edits. G. Sande (1979b) has developed a procedure for linear edits with continuous variables.

5. Miscellaneous Topics

5.1 Multiple Imputations

A serious risk with the use of imputation is that the analyst may treat the completed data set as if all the data had been collected from respondents and no imputations had been made, and hence be misled into believing that the data are more reliable than they really are. There are two dangers involved in treating the imputed values as actual values. In the first place, imputations made for missing values are dependent on the imputation model employed, whether it is the implicit model involved in assigning donors' values to recipients or whether an explicit model, such as a regression equation, is used. The analyst needs to be conscious of this model dependence and, in particular, should consider how the use of alternative models could affect the results of the analyses. Secondly, although imputation assigns values for missing responses, it does not increase the amount of data collected. Imputation does not raise the sample size from r to n responses for an item; in fact, the effective sample size may be reduced to a number less than r, the extent of the reduction depending on the level of item nonresponse and the particular imputation procedure employed (see Section 2.3.1).

Rubin (1977b, 1978, 1979a,b, 1980) has widely advocated the use of multiple imputations to reduce and quantify these dangers. On the one hand, the replication of a given imputation model can provide information on the effect of the imputation on the stability of survey results. On the other hand, the application of several different imputation models can shed light on the sensitivity of the survey results to the choice of model. A further advantage of multiple imputations, already noted at the end of Section 2.3.1, is the reduction in the variance increase caused by random imputation; see also the note by Kish (1979) included as an Appendix to this report.

In developing the theory of multiple imputations, we start with a very general result, applicable for any statistic, any sample design and any imputation procedure. Following equation (2.3.1.1), the variance of any statistic z can be expressed as

$$V(z) = E_1V_2(z) + V_1E_2(z) \quad (5.1.1)$$

where E_2 and V_2 are conditional expectation and variance over the random imputation scheme, and E_1 and V_1 are the corresponding operators for the initial sampling. The first term in (5.1.1) represents the variance from the imputation scheme. An estimator of this term can be readily generated from the use of multiple imputations, through the approach of simple replicated sampling. First, form a complete data set by applying the given imputation scheme to impute values for all missing responses. Repeat this operation independently c times, to produce c complete data sets. Compute the statistic z_j for data set j, and let $z = \Sigma z_j/c$ be the average estimator based on all c data sets. Then $V_2(z) = V_2(z.)/c$, where $V_2(z.)$ is the common variance of each of the z_j's. An unbiased estimator of $V_2(z.)$ may be simply obtained from the random sample of z_j's as $v(z.) = \Sigma(z_j - z)^2/(c - 1)$. Since $E_2[v(z.)/c] = V_2(z)$, it follows that $E_1E_2[v(z.)/c] = E_1V_2(z)$, i.e. that $v(z.)/c$ is an unbiased estimator of the imputation variance.

The generality of the above result is limited only by the assumption that the replications of the imputations are conducted independently and that the overall estimator is formed as the average of the replicate values. The latter assumption may not be met in practice, for the analyst may opt for the overall estimator formed from pooling the c data sets, $\tilde{z}$, rather than the average value, z. With linear estimators $\tilde{z} = z$, but the equality does not hold for nonlinear statistics, and $\tilde{z}$ may be preferred on the grounds of a smaller bias. However, since $\tilde{z}$ and z are generally close, it is common practice with replicated sampling to approximate the variance of $\tilde{z}$ by $v(z.)/c$ (see, for instance, Kalton, 1977). The repeated replication imputation procedure described by Kish in the Appendix and the procedure of partitioning weights described at the end of Section (2.3.1) both create pooled data sets from which the estimator $\tilde{z}$ would be obtained.

The preceding discussion has been concerned with estimating the imputation variance component $E_1V_2(z)$ of the total variance of z. When imputation is used, care is also needed to obtain a suitable estimator of the sampling variance component, $V_1E_2(z)$. In order to illustrate the issues involved, we will consider the simple situation of a SRS of size n, from which m y-values are missing. These missing values are then imputed by random imputation using unrestricted sampling within a single imputation class. Multiple imputations are used. The estimator we will consider is the sample mean $\bar{y} = \Sigma\bar{y}_j/c$. Following the notation and methods of Section 2.3.1, the variance of $\bar{y}$, $V(\bar{y}) = E_1V_2(\bar{y}) + V_1E_2(\bar{y})$, may be readily obtained as follows. First note that

$$\bar{y} = \Sigma(r\bar{y}_r + m\bar{y}_{mj})/nc = \bar{r}\bar{y}_r + (\bar{m}\Sigma\bar{y}_{mj}/c). \qquad (5.1.2)$$

Then $V_2(\bar{y}) = \bar{m}^2\Sigma V_2(\bar{y}_{mj})/c^2$ with independent replications,

$= \bar{m}(r - 1)s_r^2/nrc$ with unrestricted sampling,

$\approx \bar{m}s_r^2/nc$ treating $r - 1 \approx r$. (5.1.3)

Thus

$$E_1V_2(\bar{y}) \approx \bar{m}S_r^2/nc.$$

Now $E_2(\bar{y}) = \bar{y}_r$, so that, ignoring the f.p.c. term $[1 - (r/R)]$,

$$V_1E_2(\bar{y}) \approx S_r^2/r. \qquad (5.1.4)$$

Hence

$$V(\bar{y}) \approx (S_r^2/r) + (\bar{m}S_r^2/nc). \qquad (5.1.5)$$

In this simple case a variance estimator for $V(\bar{y})$ can be easily obtained providing the respondents' values and the imputed values can be distinguished: an unbiased estimator $s_r^2 = \Sigma(y_{ri} - \bar{y}_r)^2/(r - 1)$ of S_r^2 can be calculated, and substituted in (5.1.5). There is no need to employ the replication variance estimator $v(\bar{y}.)/c$ to estimate the imputation variance - the second term in (5.1.5). Using (5.1.2), the replicated variance estimator in fact reduces to

$$v(\bar{y}.)/c = \Sigma(\bar{y}_j - \bar{y})^2/c(c - 1)$$

$$= \bar{m}^2\Sigma(\bar{y}_{mj} - \bar{y}_m)^2/c(c - 1) = \bar{m}^2 v(\bar{y}_{mj})/c.$$

Since $E_2[v(\bar{y}_{mj})] = s_r^2/m$, the conditional expectation of the replicated variance estimator is $\bar{m}s_r^2/nc$, the value that would otherwise be substituted directly in (5.1.5). In this case, the replicated variance estimator is just a less precise estimator of $E_1V_2(\bar{y})$ than would be obtained by the direct substitution of s_r^2 for S_r^2 in the second term of (5.1.5).

Suppose now, however, that the respondents' values and imputed values cannot be distinguished, a not infrequent occurrence in practice; in addition, assume that the numbers of respondents (r) and nonrespondents (m) are unknown. In this situation, the average sample variance across replicates $s^2 = \Sigma s_j^2/c$, where $s_j^2 = \Sigma(y_{ji} - \bar{y}_j)^2/(n - 1)$, provides a reasonable estimator for S_r^2 (see Section 2.3.2), but the variance in (5.1.5) cannot be estimated because r and m are unknown. The replication estimator of the imputation variance can, however, still be computed, so that the main problem concerns the estimation of the sampling variance (S_r^2/r). Without knowledge of the extent of missing data, the analyst might be tempted to estimate the sampling variance by (s^2/n), but this estimator effectively treats the imputed values as actual values and hence overstates the precision achieved.

Rubin (1979b) has proposed the following general estimator of the variance of a statistic z with multiple imputations, an estimator which does not require knowledge of which values are actual ones and which are imputed,

$$v(z) = (\Sigma\hat{v}_j/c) + \Sigma(z_j - \bar{z})^2/(c - 1) \qquad (5.1.6)$$

where $\hat{v}_j$ is the variance of z_j obtained by treating both imputed and real values as real values. Applying this estimator in the current case gives

$$v = (\Sigma s_j^2/nc) + [v(\bar{y}.)]. \qquad (5.1.7)$$

From Section (2.3.2) $E(s_j^2) \simeq S_r^2$, and from above $E[v(\bar{y}.)] = \bar{m}S_r^2/n$, so that

$$E(v) \simeq S_r^2(1 + \bar{m})/n.$$

Thus

$$(S_r^2/r) - E(v) = \bar{m}^2 S_r^2/r > 0.$$

This result shows that v underestimates the sampling variance of $\bar{y}$, the first term in (5.1.5), let alone the total variance.

To counteract this bias, Rubin (1979) has proposed a modification to the multiple imputation procedure, developing the modification from a Bayesian perspective. We will attempt to explain it from a different viewpoint. First, we recall the missing at random assumption made about the respondents and nonrespondents, an assumption which implies that $\bar{Y}_m \simeq \bar{Y}_r$. Not knowing $\bar{y}_m$, the best estimator of $\bar{Y}$ under this assumption (i.e. the one with smallest variance) is $\bar{y}_r$, which can be generated by carrying out the imputations so that $\bar{y}_m = \bar{y}_r$. The condition $\bar{y}_m = \bar{y}_r$ can be achieved by imputing $\bar{y}_r$ for all missing values, but, as has been observed, this procedure distorts the distribution of the y-variable. A random imputation procedure assigning donors' values to recipients avoids this distortion, but at the cost of increasing the variance of the mean. The variance increase results from the relaxation of the condition that $\bar{y}_m = \bar{y}_r$ to one that $E_2(\bar{y}_m) = \bar{y}_r$. A further relaxation of the condition to $E_2(\bar{y}_m) = \bar{Y}_r$ can be used to assist with the variance estimation problem under discussion.

The condition $E_2(\bar{y}_m) = \bar{Y}_r$ can be expressed as

$$E_2(\bar{y}_m) = \bar{y}_r + (\bar{Y}_r - \bar{y}_r) = \bar{y}_r + d.$$

The value of d is of course unknown for a particular sample, but its sampling distribution is known to be normal with a mean of zero and variance of S_r^2/r, i.e. $N(0,S_r^2/r)$. Estimating S_r^2 by s_r^2, Rubin's modification is then to draw a value of d σ from

$N(0,s_r^2/r)$ and add this value, d_j, to all imputed values in replicate j. This modification changes the overall sample mean over the c replicates from $\bar{y}$ in (5.1.2) to

$$\bar{y}^* = \Sigma(\bar{r}\bar{y}_r + \bar{m}\bar{y}_{mj} + \bar{m}d_j)/c$$

$$= \Sigma(\bar{r}\bar{y}_r + \bar{m}\bar{y}_{mj})/c + \bar{m}\Sigma d_j/c. \qquad (5.1.8)$$

To establish that $\bar{y}^*$ is an unbiased estimator of $\bar{Y}_r$, we introduce another level of expectation, E_3, denoting expectation over random sampling of d_j. Since $E_3(d_j) = 0$, $E_3(\bar{y}^*) = \bar{y}$. Hence $E(\bar{y}^*) = \bar{Y}_r$.

The variance of $\bar{y}^*$ may be derived from

$$V(\bar{y}^*) = V_1E_2E_3(\bar{y}^*) + E_1V_2E_3(\bar{y}^*) + E_1E_2V_3(\bar{y}^*) \qquad (5.1.9)$$

$$= V(\bar{y}) + E_1E_2V_3(\bar{y}^*)$$

where the second term represents the loss in precision in using the estimator $\bar{y}^*$ rather than $\bar{y}$. From (5.1.8), approximating $(r - 1) \simeq r$,

$$V_3(\bar{y}^*) \simeq \bar{m}^2 s_r^2/rc \qquad (5.1.10)$$

and

$$E_1E_2V_3(\bar{y}^*) = \bar{m}^2 S_r^2/cr. \qquad (5.1.11)$$

Thus, using (5.1.5)

$$V(\bar{y}^*) \simeq (S_r^2/r) + (\bar{m}S_r^2/nc) + (\bar{m}^2 S_r^2/rc) \qquad (5.1.12)$$

$$\simeq (S_r^2/r) + (\bar{m}S_r^2/nc)[1 + (m/r)] \qquad (5.1.13)$$

$$\simeq (S_r^2/r) + (\bar{m}S_r^2/rc) \qquad (5.1.14)$$

The effect of the modification to the imputation procedure is thus to increase the imputation variance by a factor of n/r (c.f. equation (5.1.5)).

Turning to the issue of estimating $V(\bar{y}^*)$, we will first establish that $s^2 = \Sigma s_j^2/c$ provides a reasonable estimator of S_r^2. Consider $(n - 1)s_j^2 = \Sigma(y_{ji}^* - \bar{y}_j^*)^2$.

$$\Sigma^n(y_{ji}^* - \bar{y}_j^*)^2 = \Sigma^r(y_{ri} - \bar{y}_r)^2 + \Sigma^m(y_{mji}^* - \bar{y}_r)^2 - n(\bar{y}_j^* - \bar{y}_r)^2$$

$$= \Sigma^r(y_{ri} - \bar{y}_r)^2 + \Sigma^m(y_{mi} + d_j - \bar{y}_r)^2 - n(\bar{y}_j^* - \bar{y}_r)^2.$$

Now $\bar{y}_j^* - \bar{y}_r = m(d_j + \bar{y}_m - \bar{y}_r)/n$, so that, with $w_i = (1 + g_i)$ where g_i is the number of times the ith respondent is used as a donor,

$$\Sigma(y_{ji}^* - \bar{y}_j^*)^2 = \Sigma^r w_i(y_{ri} - \bar{y}_r)^2 + md_j^2 + 2md_j(\bar{y}_m - \bar{y}_r)$$
$$- (m^2 d_j^2/n) - [2m^2 d_j(\bar{y}_m - \bar{y}_r)/n]$$
$$- [m^2(\bar{y}_m - \bar{y}_r)^2/n].$$

Hence,

$$E_3\Sigma(y_{ji}^* - \bar{y}_j^*)^2 = \Sigma^r w_i(y_{ri} - \bar{y}_r)^2 + (ms_r^2/r) - (m^2 s_r^2/rn)$$
$$- [m^2(\bar{y}_m - \bar{y}_r)^2/n]$$

and

$$E_2E_3\Sigma(y_{ji}^* - \bar{y}_j^*)^2 = [n(r - 1)s_r^2/r] + [ms_r^2(1 - \bar{m})/r]$$
$$- [m^2(r - 1)s_r^2/nrm]$$
$$= [s_r^2(r - 1)(n - \bar{m})/r] + \bar{m}s_r^2$$
$$= s_r^2[n - (n/r) + (\bar{m}/r)] \qquad (5.1.15)$$
$$\simeq (n - 1)s_r^2 \text{ if } r \text{ is large.}$$

Thus $E_2E_3(s_j^2) \simeq s_r^2$, and hence $E(s^2) \simeq S_r^2$.

Now consider the replication variance estimator $v(\bar{y}_\cdot^*)/c = \Sigma(\bar{y}_j^* - \bar{y}^*)^2/c(c - 1)$. Being a replication estimator it is unbiased for the imputation variance, i.e. from (5.1.14)

$$E[v(\bar{y}^*_\cdot)/c] = \bar{m}S_r^2/rc$$

so that

$$E[v(\bar{y}^*_\cdot)] = \bar{m}S_r^2/r.$$

Combining this result with $E(s_j^2) \simeq S_r^2$ gives the expectation of $v^* = (\Sigma s_j^2/nc) + [v(\bar{y}^*_\cdot)]$ - the modified version of v from (5.1.7) - as

$$E(v^*) = (S_r^2/n) + (mS_r^2/nr) = S_r^2/r. \qquad (5.1.16)$$

Comparing $E(v^*)$ with $V(\bar{y}^*)$ in (5.1.14) shows that v^* is an unbiased estimator of the sampling variance. Providing the imputation variance is small relative to the sampling variance, as will occur when response rates are high and/or when the number of multiple imputations is sizeable, v^* will provide a reasonable estimator of $V(\bar{y}^*)$. In other cases, the additional term of $v(\bar{y}^*_\cdot)/c$ may be added to reflect the imputation variance. If this is done the variance estimator becomes

$$v' = (s^2/n) + [(c + 1)v(\bar{y}^*_\cdot)/c] \qquad (5.1.17)$$

with $E(v') = V(\bar{y}^*)$.

An alternative approach is to increase the variance of the d_j to make v^* an unbiased estimator of the variance of $\bar{y}$. Let $V(d_j) = kS_r^2/r$. Then $V_3(\bar{y}^*)$ in (5.1.10) becomes $V_3(\bar{y}_1^*) = (\bar{m}^2kS_r^2/rc)$ for the revised estimator $\bar{y}_1^*$. Hence

$$E_1E_2V_3(\bar{y}_1^*) = \bar{m}^2kS_r^2/rc \qquad (5.1.18)$$

and

$$V(\bar{y}_1^*) = (S_r^2/r) + \bar{m}S_r^2(\bar{r} + k\bar{m})/rc. \qquad (5.1.19)$$

Now $E_2E_3\Sigma(y^*_{ji} - \bar{y}^*_j)^2$ in (5.1.15) becomes

$$E_2E_3\Sigma(y^*_{ji} - y^*_j)^2 = s^2_r[n - (n/r) + \bar{m}/r + \bar{m}(k - 1)] \qquad (5.1.20)$$

which remains approximately equal to $(n - 1)s^2_r$ for large r, providing k is small. Hence, under this proviso, s^2 is still a reasonable estimator of S^2_r. The expected value of v^* is then

$$E(v^*) \simeq (S^2_r/n) + \bar{m}S^2_r(\bar{r} + k\bar{m})/r \qquad (5.1.21)$$

This is equal to $V(\bar{y}^*_1)$ in (5.1.19) if

$$(1/r) - (1/n) = \bar{m}(\bar{r} + k\bar{m})(c - 1)/rc$$

i.e. if $$k = 1 + [1/\bar{m}(c - 1)]. \qquad (5.1.22)$$

With the smallest possible value of $c = 2$, $k = 1 + (1/\bar{m})$, in which case the last term in (5.1.20) is $\bar{m}(k - 1) = \bar{m}$. Hence k is sufficiently small to make s^2 a reasonable estimator of S^2_r. Thus, if the d_j are chosen from $N(0,\sigma^2)$, where

$$\sigma^2 = (S^2_r/r)\{1 + [1/\bar{m}(c - 1)]\},$$

the estimator v^* is approximately unbiased for $V(\bar{y}^*_1)$.

The reason for incorporating the d_j term into the imputation procedure is to provide a variance estimator v^* which reflects the true variance of the sample mean $\bar{y}^*$. In deciding whether to include the d_j's, this benefit has to be balanced against the loss of precision in $\bar{y}^*$ given by (5.1.11) or the larger loss with the amended modification given by (5.1.18). In the case of Rubin's modification, the variance is increased over the simple multiple imputation variance, $V(\bar{y})$ in (5.1.5), by the factor

$$\begin{aligned} V(\bar{y}^*)/V(\bar{y}) &= [(S^2_r/r) + (\bar{m}S^2_r/rc)]/[(S^2_r/r) + (\bar{m}S^2_r/nc)] \\ &= [1 + (\bar{m}/c)]/[1 + (\bar{m}\bar{r}/c)] \\ &= n(nc + m)/(n^2c + rm) \\ &= 1 + [\bar{m}^2/(c + \bar{m}\bar{r})] \qquad (5.1.23) \end{aligned}$$

Thus, for instance, with $c = 2$ and $\bar{m} = 0.3$, the factor is 1.04, a 4 per cent increase in the variance of the sample mean. The larger the value of c, the smaller this factor is.

With the amended modification $V(\bar{y}_1^*)$ from (5.1.19) becomes

$$V(\bar{y}_1^*) = (S_r^2/r)\{1 + [\bar{m}/(c - 1)]\}$$

so that

$$\begin{aligned} V(\bar{y}_1^*)/V(\bar{y}^*) &= \{1 + [\bar{m}/(c - 1)]\}/[1 + (\bar{m}/c)] \\ &= 1 + [\bar{m}/(c - 1)(c + \bar{m})] \end{aligned} \qquad (5.1.24)$$

With $c = 2$ and $\bar{m} = 0.3$, the variance of the mean with the amended modification is thus increased by a substantial factor of 1.13 over that of the mean with the original modification.

A decision to be made with any multiple imputation procedure is the number of replications, c, to employ. The larger the value of c, the smaller is the imputation variance, and hence the greater is the precision of the sample mean. Rubin (1979) shows that, with unrestricted sampling of donors, the choice of $c = 2$ goes a good way towards removing the imputation variance. With a SRS of donors, the imputation variance with two replications is even smaller (see the Appendix), and it would be smaller still if a proportionate stratified sample of donors - as discussed in Section 2.3.1 - were used (see also Kalton and Kish, 1981). Thus, with regard to the reduction of the imputation variance, most of the gain of multiple imputations can be achieved from a small number of replications, even the minimum of 2.

When multiple imputations are used for the purpose of variance estimation, the issue of the precision of the variance estimate arises. The replication variance estimator $[v(z.)/c]$ of the conditional imputation variance $V_2(z)$ discussed earlier has $(c - 1)$ degrees of freedom. A sizeable number of degrees of freedom (say 20 to 30 or more) is needed if this estimator is to be of adequate precision for analysis purposes. The argument here corresponds to the usual one concerning the precision of replication variance estimators (see Kish, 1965,

p. 131). The concern for precision also applies to the overall variance estimator for z, v(z) given in (5.1.6).

Practical considerations also play an important role in determining the choice of c, and indeed in whether multiple imputations should be employed at all. The added complexity and cost of analyzing the data sets resulting from multiple imputations are major hurdles that need to be cleared. The larger the number of replications the larger the total data set and hence the more costly the analyses.

The total data set for multiple imputations has been described above as a set of c independently completed data sets, with any statistic of interest being computed as the average $\bar{z}$ of the c replicate values, z_j. For economy of processing the total data might be better constructed along the lines suggested in Chapter 2 of splitting records with missing data into c parts. With this scheme, the total data set comprises the records with no missing data and c parts of each of the records with missing data, each part being assigned a weight of 1/c (or w_i/c if record i has a weight of w_i). Each part of a record with missing data is given an index (1,2...c) for variance estimation purposes. The z_j's required for the computation of v(z.) are obtained by confining calculations to the records with no missing data and the jth part of split records, with the latter being given the total weight of their original records. Since the records with no missing data are included in the calculation of all the z_j's, computing economies that calculate their contribution to the z_j's only once can be readily envisaged for many types of statistics.

Although this alternative construction of the data set permits the calculation of the overall estimate as $\bar{z}$, the more natural estimator to use is $\tilde{z}$, the single estimator obtained from the total data set. The use of $\tilde{z}$ avoids the need to compute the z_j's and for reasons of bias and variance is often preferred to $\bar{z}$. Since the variance estimator used for $\tilde{z}$ is the same as that for $\bar{z}$, and this depends on z_j, it is of course true that the z_j's have to be calculated for any statistic for which a variance estimate is wanted. However, in complex surveys the number of estimates made exceeds by manyfold the number of variance estimates computed, and the z_j's need not be calculated for estimates whose variance estimates are not required.

The economy achieved through using this alternative construction of the data set depends on the extent and pattern of missing data. If the missing data are concentrated in a few records, the increase to the size of the data set by splitting these records into c parts may be relatively minor. If, however, the missing data are spread across most of the records, with say some records having missing responses for certain items and other records having missing responses for others, the size of the data set will increase substantially. When the situation begins to approach that in which all records have some missing data, it may be simplest to split all records into c parts, and impute for any missing data as necessary; in this case the size of the data set is increased c-fold.

To make the z_j's independent, as required by the theory of replicated sampling, the imputations have been described above as being carried out independently from one replication to the next. From the viewpoint of precision of the estimator ($\bar{z}$ or $\tilde{z}$), however, it would be preferable to assign the imputed values by some nonindependent procedure, say using stratified or systematic sampling for all c parts simultaneously. This is the standard conflict with replicated sampling: should one opt for an unbiased variance estimator of an estimate with larger variance, or for an upwardly biased variance estimator of an estimate with smaller variance? There is in general no obvious answer to this question. In the present case, however, since the use of even a small number of replications effectively removes most of the imputation variance, replications with independent imputations should generally serve well.

5.2 Weighted Data

Until this point, this report has assumed that the compensations for missing data are made to records which all have the same weight. In the case of weighting adjustments for unit nonresponse, the sampled units were assumed to be selected with equal probabilities. When units are selected with unequal probabilities, selection weights proportional to the inverse of the selection probabilities are needed in the analysis. In this section we consider the effect of unequal selection weights on nonresponse weighting adjustments. In the case of imputation for item nonresponse, the sample units may have weights both because of unequal selection probabilities and because of weighting adjustments made for unit nonresponse. We will discuss how these weights might affect the method of imputation used. Little attention has been given to the effects of weights on missing data compensation procedures; more research in this area seems needed.

Consider first the effect of selection weights on nonresponse weighting adjustments. The usual extension to procedures already described to take account of selection weights is to reweight according to the sums of selection weights rather than sample sizes. As an illustration, consider the estimator of the population mean with sample weighting adjustment for nonresponse. Let q_{rhi} be the selection weight for respondent i in weighting class h, and let $q_{rh} = \Sigma q_{rhi}$; let q_{mhi} and q_{mh} be the equivalent quantities for the nonrespondents; and let $q_h = q_{rh} + q_{mh}$. The adjusted estimator of the population mean is then $\bar{y}_s = \Sigma q_h \bar{y}_{rh} / \Sigma q_h$, where $\bar{y}_{rh} = \Sigma q_{rhi} y_{rhi} / \Sigma q_{rh}$. This may be expressed as $\bar{y}_s = \Sigma\Sigma w_{hi} y_{rhi} / \Sigma\Sigma w_{hi}$ with $w_{hi} \propto q_{rhi}(q_h/q_{rh})$. In other words, the selection weight is increased by the factor q_h/q_{rh} to compensate for the nonrespondents. This estimator is a straightforward generalization of the estimator with sample weighting adjustments discussed in Section 3.3. In the epsem case $q_{rhi} = q_{mhi} = 1$ for all sampled units, so that w_{hi} reduces to $w_{hi} \propto n_h/r_h$.

In general the final weights assigned, i.e. the combined selection weights and sample weighting adjustments (w_{hi}), will vary between respondents within a weighting class. There is, however, a special case when all sampled units in a weighting class have the same selection weights, say $q_{h.}$; in this case $w_{hi} \propto q_{h.}(n_h/r_h)$. This situation would, for instance, occur with a disproportionate stratified sample in which the strata are used in forming weighting classes. For example suppose that a certain subgroup of the population (e.g. the elderly) were sampled from a separate sampling frame with a higher sampling fraction than used for sampling the rest of the population, with epsem sampling within the two strata. If weighting classes were formed so that no class contained both persons sampled from the elderly stratum and persons sampled from the other stratum, then the condition for the equality of final weights for respondents within weighting classes would hold.

It may be noted that in this special case the nonresponse sample weight adjustment q_h/q_{rh} reduces to n_h/r_h. In general, the equality of q_h/q_{rh} and n_h/r_h will not hold exactly. However, if the assumption of data missing at random within the weighting class in valid, the equality should hold approximately, that is within random chance fluctuations. This may be demonstrated by considering the q-variable as simply another survey variable. We will condition on the selected sample units in the weighting class and the number of respondents, and assume that data are missing at random. Then with q_{rh}/r_h being the respondent mean weight and q_{mh}/m_h being the nonrespondent weight, it follows from the missing-at-random assumption that

$$E_2(q_{rh}/r_h) = E_2(q_{mh}/m_h) = E_2(q_h/n_h)$$

i.e. $$E_2(q_h)/E_2(q_{rh}) = n_h/r_h.$$

The implication of this result is that a marked difference between q_h/q_{rh} and n_h/r_h suggests that the missing-at-random assumption may be violated. Checks should be made on this difference, and a general comparison made between the selection weights of respondents and nonrespondents within weighting classes, to examine the appropriateness of the missing-at-random assumption. If marked differences are found, consideration needs to be given to incorporating the information reflected in the selection weights into the construction of the weighting classes. If selection weights are related to response rates and also to the survey variables, biased results will occur unless some nonresponse adjustments based on selection weights are made.

The common practice in making imputations with weighted data is to proceed as described in Chapter 4 without taking account of the weights. Little work has been done to examine the appropriateness of this procedure, but two observations can be made. First, it is apparent that there is a danger of increased imputation variance if a sizeable variability in selection weights is ignored. For example, recalling the equivalence between imputation and weighting adjustments for estimating a mean or total, imputing a donor's value to a nonrespondent with a large weight is equivalent to increasing the donor's weight by that large weight. A simple method for avoiding this large weight increase would be to partition the records and weights of recipients with sizeable weights, and make separate imputations for the parts along the lines already described at the end of Section 2.3.1 and earlier in this chapter. The added feature here is that the recipient records are partitioned into variable numbers of parts depending on the magnitudes of their weights. A possible elaboration on this method would be to partition both respondents' and nonrespondents' records into multiple parts, each part having the same weight across all sampled units. The equality of weights for all parts can be achieved by fixing the number of parts for a sampled unit proportional to its weight; thus, for instance, with integer weights, units with weights 2, 3, 5 could be divided into 2, 3 and 5 parts respectively. Imputations could then be carried out with the equal-weighted parts. This procedure bears some similarity to the probability-proportional-to-weight scheme of Cox (1980) as described below, except that in her scheme each recipient is assigned a single donor's value whereas here the several parts of a recipient's record may be assigned from different donors.

The second general observation to be made about imputations with weighted data is that, as discussed above in the case of weighting adjustments, the weights may be significantly related both to the pattern of missing data and to the survey variables. If so, it may be advisable to incorporate them in some way into the control variables used for forming the imputation classes; such a procedure would in effect be matching recipients and donors in terms of their weights.

Suppose now that imputation classes are formed in a way which leaves variability in the weights within classes, and assume that the data are missing at random within classes. A consequence of this assumption is that the class population mean of respondents is equal to that of nonrespondents, $\bar{Y}_r = \bar{Y}_m$ or $E_\xi(\bar{Y}_r) = E_\xi(\bar{Y}_m)$ as discussed in Section 2.1 (we will drop the imputation class subscript h throughout this discussion). Following the line of argument used in Chapter 2, we may then seek an imputation procedure satisfying the condition $E_2(\bar{y}_m) = \bar{y}_r$, where E_2 is the expectation over repeated applications of the imputation procedure. Cox (1980) has devised a procedure that satisfies this condition with weighted data.

We will use the following notation to describe a simplified version of the procedure. Let w_{ri} and y_{ri} be the weight and y-value for respondent i, with $w^*_{ri} = w_{ri}/\Sigma w_{ri}$; similarly for recipient i, let w_{mi} be the weight and y_{mi} be the imputed value, with $w^*_{mi} = w_{mi}/\Sigma w_{mi}$. The respondent mean is then $\bar{y}_r = \Sigma w^*_{ri} y_{ri}$ and the recipient mean is $\bar{y}_m = \Sigma w^*_{mi} y_{mi}$. Cox's procedure resembles a PPS sampling scheme from cumulative totals of the respondents' weights, but taking into account the recipients' weights as well. It is perhaps most simply explained in terms of the diagram on the next page. In the diagram the seven respondents and five recipients are listed in two columns, the columns being split up according to the weights of the units. The column of recipients' weights is mapped into the respondents' column as indicated by the dotted lines, and a recipient's imputed value is taken from his part of the respondents' column. For instance, recipient 1's imputed value comes from either respondent 1 or 2, while recipient 2's value comes from one of respondents 2, 3, 4 or 5. The weight for respondent 2 is divided into $w^*_{r2.1}$ and $w^*_{r2.2}$, the first part being associated with recipient 1 and the second part with recipient 2. Similarly, the weight for respondent 5 is divided between recipients 2 and 3. The choice between respondents 1 and 2 as donor of the imputed value for recipient 1 is made by choosing a number at random up to $w^*_{r1} + w^*_{r2.1}$. If the number is over w^*_{r1}, respondent 2 is chosen; otherwise respondent 1 is chosen. In the same way the donor for recipient 2 is chosen by drawing a random number up to $w^*_{r2.2} + w^*_{r3} + w^*_{r4} + w^*_{r5.2}$, and determining the donor from respondents 2, 3, 4 and 5 by a standard PPS selection from the

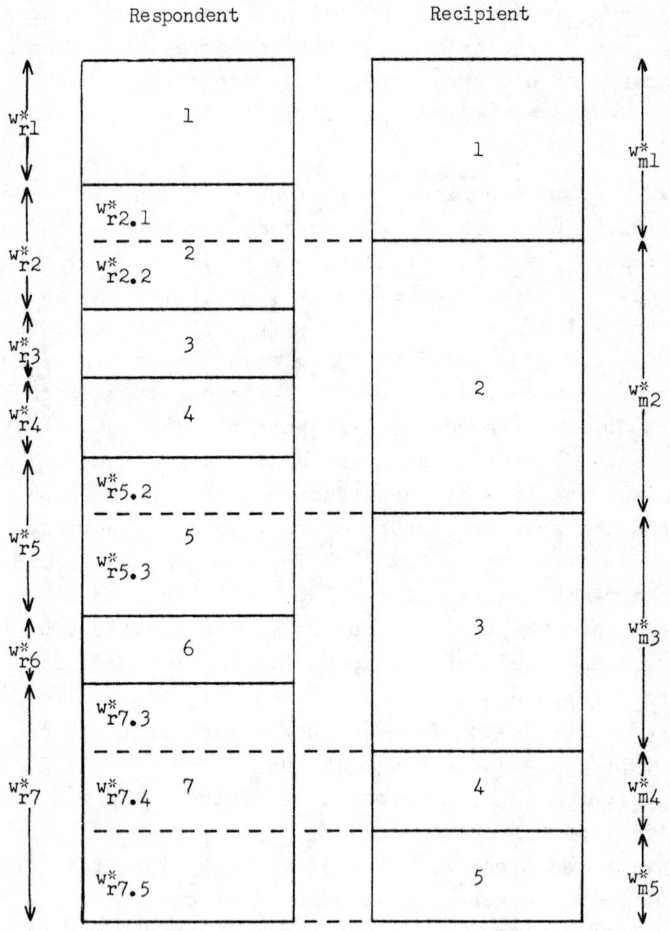

Respondent | Recipient

w^*_{r1} 1 | 1 w^*_{m1}

$w^*_{r2.1}$

w^*_{r2} 2 $w^*_{r2.2}$ | 2 w^*_{m2}

w^*_{r3} 3

w^*_{r4} 4

$w^*_{r5.2}$

w^*_{r5} 5 $w^*_{r5.3}$ | 3 w^*_{m3}

w^*_{r6} 6

$w^*_{r7.3}$

w^*_{r7} $w^*_{r7.4}$ 7 | 4 w^*_{m4}

$w^*_{r7.5}$ | 5 w^*_{m5}

cumulative totals of these values (see, for instance, Kish, 1965, Chapter 7). This procedure is applied to impute a value for each of the recipients in the class.

We will now show that this procedure satisfies the condition that $E_2(\bar{y}_m) = \bar{y}_r$. For this purpose, the recipients' mean $\Sigma^m w^*_{mj} y_{mj}$ can be expressed as

$$\bar{y}_m = \Sigma^r a_i y_{ri}$$

where $a_i = \Sigma^m_j d_{ij} w^*_{mj}$ and $d_{ij} = 1$ if respondent i donates his value to recipient j and $d_{ij} = 0$ if not. Then

$$E_2(\bar{y}_m) = \Sigma^r E_2(a_i) y_{ri}$$

and

$$E_2(a_i) = \Sigma^m E_2(d_{ij}) w^*_{mj}.$$

Now $E_2(d_{ij}|j) = P(d_{ij} = 1|j)$. In general, suppose that the mapping of recipient j's weight covers an amount $w^*_{ri.j}$ of respondent i's weight. Then

$$P(d_{ij} = 1|j) = w^*_{ri.j}/w^*_{mj}.$$

Thus

$$E_2(a_i) = \Sigma^m_j (w^*_{ri.j}/w^*_{mj}) w^*_{mj} = \Sigma^m_j w^*_{ri.j}.$$

Since all of respondent i's weight is associated with some recipient, $\Sigma^m_j w^*_{ri.j} = w^*_{ri}$. Thus $E_2(\bar{y}_m) = \Sigma^r w^*_{ri} y_{ri} = \bar{y}_r$, as required. This result is derived conditional on the particular order chosen, but it clearly also applies unconditionally.

The above procedure fairs well with regard to the avoidance of multiple uses of donors within the restrictions imposed by variable weights, but there is a slight modification which can further improve the position. As indicated in the diagram, there is a chance for a respondent mapped into two recipients to be chosen as the donor for both of them, by the choice of a high random number to determine the donor for the first recipient and a low one for the second. If a single random number between 1 and K is chosen, and scaled by a factor

w^*_{m1}/K for the first recipient and w^*_{m2}/K for the second, the use of these two scaled numbers to determine the two donors will minimize the chance of the overlap donor being selected twice. Since this modification does not affect the above theory, it remains true that $E_2(\bar{y}_m) = \bar{y}_r$. Cox's procedure, which is a more sophisticated version of that described above, minimizes the multiple use of donors.

APPENDIX

Repeated Replication Imputation Procedure (RRIP)

By: Leslie Kish

This technique obtains dramatic reductions in the increases of the variance caused by imputation for nonresponse. Like other techniques for imputations with simple randomized replications it makes several simplifying implicit assumptions:

(1) The sample is divided into two parts in which the response rates are R and $R(1 - W)$ respectively, and the two parts are treated as if they were strata.

(2) To control for (eliminate?) the differential bias due to the differential response rates the weight of the second group is increased in the ratio $1/(1 - W)$.

(3) This increase could be made with the weight $(1 - W)$ assigned to each case in the group, or by assigning the mean of the $(1 - W)$ portion to a fraction W of imputed cases, or some other way. Instead it is decided to select with simple random sampling without replacement the fraction $W/(1 - W)$ of the cases and duplicate them.

(4) The proportionate increase due to randomized duplication is the second term in the variance $1 + W(1 - W)/(1 + W)^2$.

The last term is a special case of the increase where the portion $(1 - W)$ has the weight k (integer) and portion W (SRS without replacement) has the weight $(k + 1)$. That increase in the variance has been shown (Kish, 1965, Section 11.7B, and 1976, Section 7.5) to be the second term in the ratio of SRS variances, i.e. in $1 + [W(1 - W)/(k + W)^2]$.

This ratio decreases rapidly with k. The increase has a maximum of $M = 1/4k(k + 1)$ at $\tilde{W} = k/(2k + 1)$. Values of $\tilde{W}$ and M for $k \leq 5$ are given in Table 1.

Table 1: Maximum proportionate increase in variance, $M = 1/4k(k + 1)$, at $\tilde{W} = k/(2k + 1)$.

k	1	2	3	4	5
$\tilde{W}$	1/3	2/5	3/7	4/9	5/11
M	1/8	1/24	1/48	1/80	1/120

The procedure (RRIP) advanced here is based on the rapid decreases obtainable by making $k > 1$, and k of 2, 3, 4 will do very well. The procedure is simple. We suppose that the different response portions have been identified and W_i is the proportion of duplication needed in the ith portion.

(1) First make k (2, 3 or 4) replications of the entire sample of both (or all) portions. This has no effect on simple estimates or on their variances, except for the constants of adjustment k and k^2.

(2) From <u>one</u> of the k replicates duplicate kW_i in the ith portion.

(3) If $kW_i > 1$, duplicate one (or more) entire replicates plus $(kW_i - r) < 1$ duplicates from the next replicate, with $r = 1$ or 2 or 3... .

The variance ratio is

$$1 + [kW(1 - kW)/(k + kW)^2] = 1 + [W(1 - W)/(1 + W)^2] - W[(k - 1)/k]/(1 + W)^2$$

The second term is the increase due to simple duplication and the third term shows the reduction wrought by the RRIP procedure. Table 2 presents for a few values of W and for k = 1, 2, 3, 4 and 5 the net increase in the variance on the top and the reduction on the bottom. The net increase becomes .0000 when $kW = 1$, and is shown with a "*". When $kW > 1$, k in the formula becomes $(k + 1)$ in actuality.

Table 2: Net increase in variance (on top) and reduction due to RRIP (on bottom).

k=	1	2	3	4	5
W = .05	.0431	.0204	.0129	.0091	.0068
	0	.0227	.0302	.0340	.0363
W = .10	.0744	.0331	.0193	.0124	.0083
	0	.0413	.0551	.0620	.0661
W = .15	.0964	.0397	.0208	.0113	.0057
	0	.0567	.0756	.0851	.0907
W = .20	.1111	.0417	.0185	.0069	*.0000
	0	.0694	.0926	.1042	.1111
W = .33	.1250	.0306	*.0000	*.0078	*.0050
	0	.0944	.1250		

This RRIP technique is closely related to the multiple imputation technique introduced by Rubin (1977, 1978, 1979b). This note was stimulated by reading the large gains he shows for multiple imputation and by working on our own project. But those results coincide neatly with the drastic reduction of the variance increase from replication where the effects due to differences in weights k and (k+1) are reduced by increasing k in the denominator of $W(1 - W)/(k + W)^2$. This, in turn, is just the special case of dichotomies for the general case of $(\Sigma W_i k_i^2)(\Sigma W_i)$ for variance increase for a range (1 to K) of relative weights k_i (Kish 1965, 1976).

Both techniques depend on reducing the range of weights by averaging multiple duplications. Rubin's technique first duplicates for nonresponses and then replicates that operation. But RRIP first replicates the sample and then duplicates for nonresponse. RRIP seems to me a simpler way to obtain random selection _without_ replacement for the duplicates, and we want this in practice, I assume. Rubin's theory has duplication _with_ replacement only for simplicity; that is why his maximum increase of the variance from duplication is 0.250 (at W = 1/2) instead of 0.125 (at W = 1/3). That doubling of the variance increase is an extreme, but even for lower values of W sampling _with_ replacement within the sample size n would bring nontrivial increases in the variance, and would be opposed - correctly - in practice.

Several important relevant points must wait for later developments, but should be mentioned here. Most have been noted in passing in Kish (1965, Section 11.7, and 1976). These imputations treat total nonresponse. Item nonresponses need another treatment.

(1) In practical situations the number of groups used for imputations is much larger than two. But the principles used here for measuring gains can be extended readily to more groups.

(2) Duplicating introduces the ratio 1:2 into weights. Triplicating and replicating with higher integral weights 1:k (k = 3,4...) would increase the variance even more. Hence higher ratios and sampling with replacement should be avoided.

(3) The increases of variance are greater for larger proportions of duplicates (up to W = 1/3). It follows readily that duplication should not be to 100 per cent, but only to the group with the largest response rate to $(1 - W_{min})$.

(4) Even the above limit should be lowered somewhat (say by 0.95 or so). Our formula for k = 0 becomes $W(1 - W)/(0 + W)^2 = (1 - W)/W$, the increase of variance due to elimination of the proportion (1 - W). For small proportions this increase is only a little greater than duplicating the proportion (1 - W).

Suppose for simplicity that there are two strata only, with proportions A and (1 - A) of the entire sample, and with response rates R(1 + W) and R respectively. We can duplicate the proportion W in the second stratum to bring its weight up to R(1 + W) hence increasing the variance by $(1 - A)W(1 - W)/(1 + W)^2$. Or we can eliminate the proportion W/(1 + W) from the first stratum to bring its weight down to R. The relative increase, with (1 - W') = W/(1 + W), is $W'(1 - W')/(0 + W')^2 = W$; hence this would increase the variance by AW. The two losses stand in the ratio $F = [(1 - A)/A][(1 - W)/(1 + W)^2]$. Eliminating from A is better when A and W are small when F > 1, but duplicating from (1 - A) becomes preferable with larger W and A > 0.5 when

$F < 1$. However even better than either extreme can be the elimination of a smaller proportion C from A and duplicating for the rest (W - C) in (1 - A).

This technique is feasible also with more strata and we have applied it long ago for modest but nontrivial gains. However with RRIP the gains become larger and this refinement probably trivial.

(5) Matching cases are sometimes imputed instead of random selections from a larger group: the matching case found for each missing case is "nearest" in some (vector of) variable(s). The RRIP technique of multiple imputation may also be applied to matching to decrease the range from 1:2 to 1:k (and k = 2,3,4...) with these procedures:

(a) Replicate the response deck k times.

(b) From the original deck make k "nearest" matchings for each nonresponse.

(c) Add one of the k matchings to each of the decks.

(6) It is likely that some improvement over simple random selection of k selections can be made with some kind of control, especially with stratified sampling.

Such controls and stratification for the separate imputations for several nonresponses in the same group could further constrain the variance increase from imputations.

(7) We must note that all forms of imputations and weighting not only increase the variances but also complicate their estimation. All inferential statistics, probability statements, such as tests of significance and confidence intervals, are affected. This weighty subject must be treated separately.

In all of the above it is assumed that a sample (random or epsem) of the responses in subclasses is imputed for the nonresponses. It is likely that nonresponses differ from responses even within subclasses. To reduce those biases should be a good field for applying models, subjective and Bayesian methods, and Rubin (1977a, 1978) presents the issues well and with strength. But there are deep problems both of practical methods and of public policy, and we must avoid them in this brief note.

(8) Complex, multivariate, analytical statistics need deeper treatment.

Bibliography

Afifi, A.A. and Elashoff, R.M. (1966). Missing observations in multivariate statistics I. Review of the literature. Journal of the American Statistical Association, 61, 595-604.

Afifi, A.A. and Elashoff, R.M. (1967). Missing observations in multivariate statistics II. Point estimation in simple linear regression. Journal of the American Statistical Association, 62, 10-29.

Afifi, A.A. and Elashoff, R.M. (1969). Missing observations in multivariate statistics III. Large sample analysis of simple linear regression. Journal of the American Statistical Association, 64, 337-358.

Afifi, A.A. and Elashoff, R.M. (1969). Missing observations in multivariate statistics IV. A note on simple linear regression. Journal of the American Statistical Association, 64, 359-365.

Aigner, D.J., Goldberger, A.S. and Kalton, G. (1975). On the explanatory power of dummy variable regressions. International Economic Review, 16, 503-510.

Aigner, D.J. and Hausman, J.A. (1980). Correction for truncation bias in the analysis of experiments in time-of-day pricing of electricity. Bell Journal of Economics, 11, 131-142.

Amemiya, T. (1973). Regression analysis when the dependent variable is truncated normal. Econometrica, 41, 997-1016.

Anderson, D.W., Kish, L. and Cornell, R.G. (1980). On stratification, grouping and matching. Scandinavian Journal of Statistics, 7, 61-66.

Anderson, H. (1978). On nonresponse bias and response probabilities. Scandinavian Journal of Statistics, 6, 107-112.

Anderson, T.W. (1957). Maximum likelihood estimates for a multivariate normal distribution when some observations are missing. Journal of the American Statistical Association, 52, 200-203.

Arora, H.R. and Brackstone, G.J. (1977a). An investigation of the properties of raking ratio estimators with simple random sampling. Survey Methodology, 3, 62-83.

Arora, H.R. and Brackstone, G.J. (1977b). An investigation of the properties of raking ratio estimators: II with cluster sampling. Survey Methodology, 3, 232-252.

Ashraf, A. and Macredie, I. (1978). Edit and imputation in the Labour Force Survey. *Imputation and Editing of Faulty or Missing Survey Data*, U.S. Department of Commerce, 114-119. *Proceedings of the Section on Survey Research Methods*, *American Statistical Association*, 1978, 425-430.

Astin, A.W. and Molin, L.D. (1972). Correcting for nonresponse bias in followup surveys. Unpublished paper, Office of Research, American Council on Education, Washington, D.C.

Bailar, B.A. and Bailar III, J.C. (1979). Comparison of the biases of the "hot deck" imputation procedure with an "equal-weights" imputation procedure. *Symposium on Incomplete Data: Preliminary Proceedings* (Panel on Incomplete Data of the Committee on National Statistics/National Research Council), 422-447. U.S. Department of Health, Education, and Welfare, Washington D.C.

Bailar, B.A., Bailey, L. and Corby, C.A. (1978). A comparison of some adjustment and weighting procedures for survey data. *Survey Sampling and Measurement* (Namboodiri, N. ed.), 175-198, Academic Press, New York.

Bailar III, J.C. and Bailar, B.A. (1978). Comparison of two procedures for imputing missing survey values. *Imputation and Editing of Faulty or Missing Survey Data*, U.S. Department of Commerce, 65-75. *Proceedings of the Section on Survey Research Methods*, *American Statistical Association*, 1978, 462-467.

Banister, J. (1980). Use and abuse of census editing and imputation. *Asian and Pacific Census Forum, East-West Population Institute*, 6, No. 3.

Bankier, M.D. (1978). An estimate of the efficiency of raking ratio estimators under simple random sampling. *Survey Methodology*, 4, 115-124.

Banks, M.J. (1977). An indication of the effects of noninterview adjustment and post-stratification on estimates from a sample survey. *Proceedings of the Social Statistics Section*, *American Statistical Association*, 1977(1), 291-295.

Bartholomew, D.J. (1961). A method of allowing for 'not at home' bias in sample surveys. *Applied Statistics*, 10, 52-59.

Basu, D. (1979). A discussion on survey theory. *Symposium on Incomplete Data: Preliminary Proceedings* (Panel on Incomplete Data of the Committee on National Statistics/National Research Council), 508-513, U.S. Department of Health, Education, and Welfare, Washington, D.C.

Beale, E.M. and Little, R.J. (1975). Missing values in multivariate analysis. Journal of the Royal Statistical Society, B, 37, 129-145.

Biemer, P.P. (1980). A survey error model which includes edit and imputation error. Proceedings of the Section on Survey Research Methods, American Statistical Association, 1980, 610-615.

Bishop, Y.M.M. (1980). Imputation, revision, and seasonal adjustment. Proceedings of the Section on Survey Research Methods, American Statistical Association, 1980, 567-570.

Blumenthal, S. (1968). Multinomial sampling with partially categorized data. Journal of the American Statistical Association, 63, 542-551.

Bogeström, B., Larsson, M. and Lyberg, L. (1981). Bibliography on non-response and related topics. Swedish National Central Bureau of Statistics, Stockholm.

Brackstone, G.J. and Rao, J.N.K. (1981). An investigation of raking ratio estimators. Sankhya C., 41, 97-114.

Breiman, L. (1981). Issues in regression analysis of incomplete data. Proceedings of the Section on Statistical Computing, American Statistical Association, 1981.

Brewer, K.R. and Särndal, C.E. (1979). Six approaches to enumerative survey sampling. Symposium on Incomplete Data; Preliminary Proceedings (Panel on Incomplete Data of the Committee on National Statistics/National Research Council), 499-507. U.S. Department of Health, Education, and Welfare, Washington, D.C.

Brooks, C.A. and Bailar, B.A. (1978). An Error Profile: Employment as Measured by the Current Population Survey. Statistical Policy Working Paper 3. U.S. Department of Commerce. U.S. Government Printing Office, Washington, D.C.

Buck, S.F. (1960). A method of estimation of missing values in multivariate data suitable for use with an electronic computer. Journal of the Royal Statistical Society, B, 22, 302-306.

Cassel, C.M., Särndal, C.E. and Wretman, J.H. (1979). Some uses of statistical models in connection with the nonresponse problem. Symposium on Incomplete Data: Preliminary Proceedings (Panel on Incomplete Data of the National Statistics/National Research Council), 188-215. U.S. Department of Health, Education, and Welfare, Washington, D.C.

Chapman, D.W. (1974). An Investigation of Nonresponse Imputation Procedures for the Health and Nutrition Examination Survey. Westat, Rockville, Maryland.

Chapman, D.W. (1976). A survey of nonresponse imputation procedures. Proceedings of the Social Statistics Section, American Statistical Association, 1976(1), 245-251.

Chapman, D.W. (1979). The impact of substitution on survey estimates. Paper presented to the Panel on Missing Data: Theory Chapters, August 1979.

Cochran, W.G. (1968). The effectiveness of adjustment by subclassification in removing bias in observational studies. Biometrics, 24, 295-313.

Cochran, W.G. (3rd ed., 1977). Sampling Techniques. Wiley, New York.

Coder, J., Feldman, A. and Nelson, C. (1978). Reporting of quarterly earnings amounts in the Income Survey Development Program Site Research Sample. Imputation and Editing of Faulty or Missing Survey Data, U.S. Department of Commerce, 136-141. Proceedings of the Social Statistics Section, American Statistical Association, 1978, 627-632.

Coder, J. (1978). Income data collection and processing from the March Income Supplement to the Current Population Survey. The Survey of Income and Program Participation Proceedings of the Workshop on Data Processing, February 23-24, 1978 (D. Kasprzyk ed.), Chapter II. Income Survey Development Program, U.S. Department of Health, Education, and Welfare, Washington, D.C.

Cohen,S.B. (1981). An analysis of alternative imputation strategies for individuals with partial data in the National Medical Care Expenditure Survey. Paper presented at the 109th Annual Meetings of the American Public Health Association, Los Angeles, November, 1981.

Colledge, M.J., Johnson, J.H., Pare, R. and Sande, I.G. (1978). Large scale imputation of survey data. Imputation and Editing of Faulty or Missing Survey Data, U.S. Department of Commerce, 102-107. Proceedings of the Section on Survey Research Methods, American Statistical Association, 1978, 431-436.

Colledge, M.J., Johnson, J.H., Pare, R. and Sande, I.G. (1978). Large scale imputation of survey data. Survey Methodology, 4, 203-224.

Cox, B.G. (1980). The weighted sequential hot deck imputation procedure. Proceedings of the Section on Survey Research Methods, American Statistical Association, 1980, 721-726.

Cox, B.G. and Folsom, R.E. (1978). An empirical investigation of alternative item nonresponse adjustments. Imputation and Editing of Faulty or Missing Survey Data, U.S. Department of Commerce, 51-55. Proceedings of the Section on Survey Research Methods, American Statistical Association, 1978, 219-223.

Cox, B.G. and Folsom, R.E. (1981). An evaluation of weighted hot deck imputation for unreported health care visits. Proceedings of the Section on Survey Research Methods, American Statistical Association, 1981, 412 - 417.

Dagenais, M.G. (1971). Further suggestions concerning the utilization of incomplete observations. Journal of the American Statistical Association, 66, 93-98.

Dalenius, T. (1979). Informed consent or R.S.V.P. Symposium on Incomplete Data: Preliminary Proceedings (Panel on Incomplete Data of the Committee on National Statistics/National Research Council), 95-134. U.S. Department of Health, Education, and Welfare, Washington, D.C.

Dalenius, T. (1979). Some reflections on the problem of "missing data". Symposium on Incomplete Data: Preliminary Proceedings (Panel on Incomplete Data of the Committee on National Statistics/National Research Council), 521-523. U.S. Department of Health, Education, and Welfare, Washington, D.C.

Daniel, W.W., Schott, B., Atkins, F.C. and Davis, A. (1982). An adjustment for nonresponse in sample surveys. Educational and Psychological Measurement, 42, 57 - 67.

Dear, R.E. (1959). A principle component missing data method for multiple regression models. System Development Corporation. Report, sp-86.

Deming, W.E. (1943). Statistical Adjustment of Data. Wiley, New York.

Deming, W.E. and Stephan, F.F. (1940). On a least squares adjustment of a sample frequency table when the expected marginal totals are known. Annals of Mathematical Statistics, 11, 427-444.

Dempster, A.P., Laird, N.M. and Rubin, D.B. (1977). Maximum likelihood from incomplete data via the EM algorithm. Journal of the Royal Statistical Society, B, 39, 1-38.

DeGroot, M. and Goel, P. (1980). Estimation of the correlation coefficient from a broken random sample. Annals of Statistics, 8, No. 2, 264-278.

Drew, J.H. and Fuller, W.A. (1980). Modeling nonresponse in surveys with callbacks. Proceedings of the Section on Survey Research Methods, American Statistical Association, 1980, 639-642.

Drew, J.H. and Fuller, W.A. (1981). Nonresponse in complex multiphase surveys. Proceedings of the Section on Survey Research Methods, American Statistical Association, 1981, 623 - 628.

Dunkelberg, W.C. and Day, G.S. (1973). Nonresponse bias and callbacks in sample surveys. Journal of Marketing Research, 10, 160-168.

Edgett, G.L. (1956). Multiple regression with missing observations among the independent variables. Journal of the American Statistical Association, 51, 122-132.

Ellis, R.A., Endo, C. and Armer, M. (1970). The use of potential nonrespondents for studying nonresponse bias. Pacific Sociological Review, 13, 103-109.

Ericson, W.A. (1967). Optimal sample designs with nonresponse. Journal of the American Statistical Association, 62, 63-78.

Ernst, L.R. (1978). Weighting to adjust for partial nonresponse. Imputation and Editing of Faulty or Missing Survey Data, U.S. Department of Commerce, 76-81. Proceedings of the Section on Survey Research Methods, American Statistical Association, 1978, 468-473.

Ernst, L.R. (1980). Variance of the estimated mean for several imputation procedures. Proceedings of the Section on Survey Research Methods, American Statistical Association, 1980, 716-720.

Evans, R.W., Cooley, P.C. and Piserchia, P.V. (1979). A test for evaluating missing data imputation procedures. Proceedings of the American Statistical Association, Social Statistics Section, 1979, 469-474.

Fellegi, I.P. and Holt, D. (1976). A systematic approach to automatic edit and imputation. Journal of the American Statistical Association, 71, 17-35.

Ferrari, P.W. and Bailey, L. (1981). Preliminary results of 1980 decennial census telephone followup nonresponse experiment. Proceedings of the Section on Survey Research Methods, American Statistical Association, 1981, 264-269.

Filion, F.L. (1976). Exploring and correcting for nonresponse bias using follow-ups of nonrespondents. Pacific Sociological Review, 19, 401-408.

Folsom, R.E. (1981). The equivalence of generalized double sampling regression estimators, weight adjustments, and randomized hot deck imputations. Proceedings of the Section on Survey Research Methods, American Statistical Association, 1981, 400 - 405.

Ford, B. (1976). Missing data procedures: a comparative study. Proceedings of the Social Statistics Section, American Statistical Association, 1976, 324-329.

Ford, B. (1980). An overview of hot deck procedures. Draft paper for Panel on Incomplete Data, Committee on National Statistics, National Academy of Sciences.

Ford, B.L., Kleweno, D.G. and Tortora, R.D. (1980). A simulation study to compare procedures which impute for missing items on an ESS hog survey. Statistical Research Division, Economics and Statistics Service, U.S. Department of Agriculture, Washington, D.C.

Ford, B.L., Kleweno, D.G., and Tortora, R.D. (1980). The effects of procedures which impute for missing items: a simulation study using an agricultural survey. Proceedings of the Section on Survey Research Methods, American Statistical Association, 1980, 251-256.

Ford, B.L., Kleweno, D.G. and Tortora, R.D. (1981). The effects of procedures which impute for missing items: a simulation study using an agricultural survey. Current Topics in Survey Sampling (D. Krewski, R. Platek and J.N.K. Rao, eds.) pp. 413-436. Academic Press, New York.

Frankel, L.R. and Dutka, S. (1979). Survey design in anticipation of nonresponse and imputation. Symposium on Incomplete Data: Preliminary Proceedings (Panel on Incomplete Data of the Committee on National Statistics/National Research Council), 72-94. U.S. Department of Health, Education, and Welfare, Washington, D.C.

Freund, R. and Hartley, H.O. (1967). A procedure for automatic data editing. Journal of the American Statistical Association, 62, 341-352.

Fuchs, C. (1982). Maximum likelihood estimation and model selection in contingency tables with missing data. Journal of the American Statistical Association, 77, 270-278.

Fuller, C.H. (1974). Weighting to adjust for survey nonresponse. Public Opinion Quarterly, 38, 239-246.

Ghangurde, P.D. and Mulvihill, J. (1978). Non-response and imputation in longitudinal estimation in LFS. Unpublished paper, Household Surveys Development Staff, February 1978, Statistics Canada.

Gillo, M.W. and Shelley, M.W. (1974). Predictive modeling of multivariable and multivariate data. Journal of the American Statistical Association, 69, 646-653.

Glasser, M. (1964). Linear regression analysis with missing observations among the independent variables. Journal of the American Statistical Association, 59, 834-844.

Gleason, T. and Staelin, R. (1975). A proposal for handling missing data. Psychometrika, 40, 229-252.

Goudy, W.J. (1976). Nonresponse effects on relationships between variables. Public Opinion Quarterly, 40, 360-369.

Gower, A.R. (1979). Non-response in the Canadian Labour Force Survey. Survey Methodology, 5, 29-58.

Gray, G.B. and Platek, R. (1980). Imputation methodology. Draft paper for Panel on Incomplete Data, Committee on National Statistics, National Academy of Sciences.

Greenberg, B. (1981). Developing an edit system for industry statistics. Computer Science and Statistics: Proceedings of the 13th Symposium on the Interface, 11-16. Springer-Verlag, New York.

Greenlees, J.S., Reece, W.S. and Zieschang, K.D. (1982). Imputation of missing values when the probability of response depends on the variable being imputed. Journal of the American Statistical Association, 77, 251-261.

Griliches, Z., Hall, B.H. and Hausman, J.A. (1977). Missing data and self-selection in large panels. Presented to the INSEE Conference on "Econometrics of Panel Data," August, 1977, Paris, France.

Haitovsky, Y. (1968). Missing data in regression analysis. Journal of the Royal Statistical Society, B, 30, 67-82.

Hansen, M.H. and Hurwitz, W.N. (1946). The problem of nonresponse in sample surveys. Journal of the American Statistical Association, 41, 517-529.

Hansen, M.H., Hurwitz, W.N. and Madow, W.G. (1953). Sample Survey Methods and Theory. Volume I Methods and Application. Volume II Theory. Wiley, New York.

Hartley, H.O. and Hocking, R.R. (1971). The analysis of incomplete data. Biometrics, 27, 783-823.

Hasselblad, V., Creason, J.P. and Stead, A.G. (1979). Applications of the missing information principle. Symposium on Incomplete Data: Preliminary Proceedings (Panel on Incomplete Data of the Committee on National Statistics/National Research Council), 251-284. U.S. Department of Health, Education, and Welfare, Washington, D.C.

Hausman, J.A. and Wise, D.A. (1979). Attrition bias in experimental and panel data: the Gary Income Maintenance Experiment. Econometrica, 47, 455-473.

Hawkins, D. (1975). Estimation of nonresponse bias. Sociological Methods and Research, 3, 461-485.

Healy, M. and Westmacott, M. (1956). Experiments analyzed on automated computers. Applied Statistics, 5, 203-206.

Heckman, J.J. (1976). The common structure of statistical models of truncation, sample selection and limited dependent variables and a sample estimator for such models. Annals of Economic and Social Measurement, 5, 475-492.

Heeringa, S.G. (1980). Nonresponse in the ISDP 1978 Research Panel: An Examination of Rates and Patterns of Nonresponse in the April (ISDP303) and July (ISDP403) Interviews. Survey Research Center, University of Michigan.

Hertel, B.R. (1976). Minimizing error variance introduced by missing data routines in survey analysis. Sociological Methods and Research, 4, 459-474.

Herzog, T.N. (1980). Multiple imputation of individual Social Security amounts, Part II. Proceedings of the Section on Survey Research Methods, American Statistical Association, 1980, 404-407.

Herzog, T.N. and Lancaster, C. (1980). Multiple imputation of individual Social Security amounts, Part I. Proceedings of the Section on Survey Research Methods, American Statistical Association, 1980, 398-403.

Hill, C.J. (1978). A report on the application of a systematic method of automatic edit and imputation to the 1976 Canadian Census. Imputation and Editing of Faulty or Missing Survey Data, U.S. Department of Commerce, 82-87. Proceedings of the Section on Survey Research Methods, American Statistical Association, 1978, 474-479.

Hill, C.J. (1978). The application of a systematic method of automatic edit and imputation to the 1976 Canadian Census of Population and Housing. Survey Methodology, 4, 178-202.

Hill, C.J. and Puderer, H.A. (1981). Data adjustment procedures in the 1981 Canadian Census of Population and Housing. Current Topics in Survey Sampling (D. Krewski, R. Platek and J.N.K. Rao, eds.) pp. 437-454. Academic Press, New York.

Hill, M.A. and Dixon, W.J. (1981). Methods for detecting patterns of missingness in incomplete data. Proceedings of the Section on Statistical Computing, American Statistical Association, 1981.

Hirschberg, D., Yuskavage, R. and Scheuren, F. (undated). The impact on personal and family income of adjusting the Current Population Survey for undercoverage. Unpublished paper, Social Security Administration, Washington, D.C.

Hocking, R.R. (1979). The design and analysis of sample surveys with incomplete data: reduction of respondent burden. Symposium on Incomplete Data: Preliminary Proceedings) (Panel on Incomplete Data of the Committee on National Statistics/National Research Council), 135-163. U.S. Department of Health, Education, and Welfare, Washington, D.C.

Hocking, R.R., Huddleston, H.F. and Hunt, H.H. (1974). A procedure for editing survey data. Applied Statistics, 23, 121-133.

Hocking, R.R. and Marx, D.L. (1979). Estimation with incomplete data: an improved computational method and the analysis of nested data. Communications in Statistics, Theory and Methods, A8(12), 1155-1181.

Hocking, R.R. and Oxspring, H.H. (1971). Maximum likelihood estimation with incomplete multinomial data. Journal of the American Statistical Association, 66, 65-70.

Hocking, R.R. and Oxspring, H.H. (1974). The analysis of partially categorized contingency data. Biometrics, 30, 469-483.

Hocking, R.R. and Smith, W.B. (1968). Estimation of parameters in the multivariate normal distribution with missing observations. Journal of the American Statistical Association, 63, 159-173.

Huddleston, H.F. and Hocking, R.R. (1978). Imputation in agricultural surveys. Imputation and Editing of Faulty or Missing Survey Data, U.S. Department of Commerce, 88-93. Proceedings of the Section on Survey Research Methods, American Statistical Association, 1978, 480-485.

Jackson, E.C. (1968). Missing values in linear multiple discriminant analysis. Biometrics, 24, 835-844.

Jones, R.G. (1979). An examination of methods of adjusting for nonresponse to a mail survey: a mail-interview comparison. Symposium on Incomplete Data: Preliminary Proceedings (Panel on Incomplete Data of the Committee on National Statistics/National Research Council), 388-409. U.S. Department of Health, Education, and Welfare, Washington, D.C.

Kalsbeek, W.D. (1980). A conceptual review of survey error due to nonresponse. Proceedings of the Section on Survey Research Methods, American Statistical Association, 1980, 131-136.

Kalton, G. (1977). Practical methods for estimating sampling errors. Bulletin of the International Statistical Institute, 47(3), 495-514.

Kalton, G., Kasprzyk, D. and Santos, R. (1980). Some problems of nonresponse and nonresponse adjustment in the Survey of Income and Program Participation. Proceedings of the Section on Survey Research Methods, American Statistical Association, 1980, 501-506.

Kalton, G., Kasprzyk, D. and Santos, R. (1981). Issues of nonresponse and imputation in the Survey of Income and Program Participation. Current Topic in Survey Sampling. (D. Krewski, R. Platek and J.N.K. Rao, eds.) pp.455-480. Academic Press, New York.

Kalton, G. and Kish, L. (1981). Two efficient random imputation procedures. Proceedings of the Section on Survey Research Methods, American Statistical Association, 1981, 146 - 151.

Kasmel, (undated). Effect on variance of adjusting for nonresponse by hot decking. Australian Bureau of Statistics.

Kaufman, G.M. and King, B. (1973). A Bayesian analysis of nonresponse in dichotomous processes. Journal of the American Statistical Association, 68, 670-678.

Kelejian, H.H. (1969). Missing observations in multivariate regression: efficiency of a first order. Journal of the American Statistical Association, 64, 1609-1616.

Kemsley, W. (1976). National Food Survey - a study of differential response based on comparison of the 1971 sample with the Census. Statistical News, 35, 18-22.

Kim, J. and Curry, J. (1977). The treatment of missing data in multivariate analysis. Sociological Methods and Research, 6, No. 2, 215-240.

King, B.F. (1980). Quota sampling. Draft paper for Panel on Incomplete Data, Committee on National Statistics, National Academy of Sciences.

Kish, L. (1965). Survey Sampling. Wiley, New York.

Kish, L. (1976). Optima and proxima in linear sample designs. Journal of the Royal Statistical Society, A, 139, 80-95.

Kish, L. (1979). Repeated Replication Imputation Procedure (RRIP). Unpublished paper, Survey Research Center, University of Michigan, Ann Arbor. (Reproduced as an Appendix to this report.)

Kish, L. (1980). Diverse adjustments for missing data. Proceedings of the 1980 Conference on Census Undercount, 83-87, U.S. Bureau of the Census, Washington, D.C.

Kish, L. and Anderson, D.W. (1978). Multivariate and multipurpose stratification. Journal of the American Statistical Association, 73, 24-34.

Koch, G.G., Imrey, P.B. and Reinfurt, D.W. (1972). Linear model analysis of categorical data with incomplete response vectors. Biometrics, 28, 663-692.

Konijn, H.S. (1981). Biases, variances and covariances of raking ratio estimators for marginal and cell totals and averages of observed characteristics. Metrika, 28, 109-121.

Koopman, R.F. (1976). Fast regression estimates of missing data. Psychometrika, 41, 277.

Kulka, R.A., Shirey, S.J., Moore, R.P. and Woodbury, N. (1981). A factorial experiment on the responses of professional nurses to a national mail survey. Proceedings of the Section on Survey Research Methods, American Statistical Association, 1981, 270-275.

Lagay, B.W. (1969). Assessing bias: a comparison of two methods. Public Opinion Quarterly, 33, 615-618.

Lansing, J.B. and Eapen, A.T. (1959). Dealing with missing information in surveys. Journal of Marketing, 24, 21-27.

Lessler, J.T. (1979). An expanded survey error model. Symposium on Incomplete Data: Preliminary Proceedings (Panel on Incomplete Data of the Committee on National Statistics/National Research Council), 371-387. U.S. Department of Health, Education, and Welfare, Washington, D.C.

Lindstrom, H., et al. (1979). Standard Methods for Non-response Treatment in Statistical Estimation. National Central Bureau of Statistics, Sweden.

Lininger, C.A. (1980). The goals and objectives of the Survey of Income and Program Participation. Proceedings of the Section on Survey Research Methods, American Statistical Association, 1980, 480-485.

Little, R.J.A. (1979). Maximum likelihood inference for multiple regression with missing values: a simulation study. Journal of the Royal Statistical Society, B, 41, 76-87.

Little, R.J.A. (1982). Models for nonresponse in sample surveys. Journal of the American Statistical Association, 77, 237-250.

Little, R.J.A. and Rubin, D.B. (1979). Six approaches to enumerative survey sampling. Discussion. Symposium on Incomplete Data: Preliminary Proceedings (Panel on Incomplete Data of the Committee on National Statistics/National Research Council), 515-520. U.S. Department of Health, Education, and Welfare, Washington, D.C.

Lord, F.M. (1955). Estimation of parameters from incomplete data. Journal of the American Statistical Association, 50, 870-876.

Lyberg, L. and Rapaport, E. (1979). Nonresponse problems at the National Central Bureau of Statistics. Unpublished paper, National Central Bureau of Statistics, Sweden.

Madow, W.G. and Rizvi, M.H. (1979). On incomplete data: a review. Draft paper for Panel on Incomplete Data, Committee on National Statistics, National Academy of Sciences.

Marini, M.M., Olsen, A.R. and Rubin, D.B. (1979). Maximum-likelihood estimation in panel studies with missing data. Sociological Methodology 1980, (K. Schuessler ed.), Chapter 11. Jossey-Bass, San Francisco.

Mathematica Policy Research (1979). Survey of Income and Program Participation, site test analysis: analysis of household and person observation losses from the site research test panel. Unpublished paper, Mathematica Policy Research Inc., Washington, D.C.

Matthai, A. (1951). Estimation of parameters from incomplete data with application to design of sample surveys. Sankhya, 11, 145-152.

Moeller, J. (undated). Expenditure regression analysis and imputation procedures with the 1972/1973 Consumer Expenditure Survey Data. Unpublished paper, Mathematica Policy Research, Inc., Washington, D.C.

Morris, C.N. (1979). Nonresponse issues in public policy experiments, with emphasis on the Health Insurance Study. Symposium on Incomplete Data: Preliminary Proceedings (Panel on Incomplete Data of the Committee on National Statistics/National Research Council), 448-470. U.S. Department of Health, Education, and Welfare, Washington, D.C.

Mulvihill, J. and Lawes, M. (1980). Imputation procedures for LFS longitudinal files. Memorandum to M.P. Singh dated March 18, 1980, Statistics Canada.

Murthy, M.N. (1979). A framework for studying incomplete data with a reference to the experience in some countries of Asia and the Pacific. Symposium on Incomplete Data: Preliminary Proceedings (Panel on Incomplete Data of the Committee on National Statistics/National Research Council), 1-20. U.S. Department of Health, Education, and Welfare, Washington, D.C.

Nathan, G. (1981). Regression analysis under differential non-response. Proceedings of the Section on Survey Research Methods, American Statistical Association, 1981, 618 - 622.

Naus, J.I. (1975). Data Quality Control and Editing. Marcel Dekker, New York.

Nicholson Jr., G.E. (1957). Estimation of parameters from incomplete multivariate samples. Journal of the American Statistical Association, 52, 523-526.

Nordbotten, S. (1963). Automatic editing of individual statistical observations. Statistical Studies, 2.

Nordbotten, S. (1965). The efficiency of automatic detection and correction of errors in individual observations as compared with other means of improving the quality of statistics. Bulletin of International Statistical Institute, 4(1), 442-472.

Nordheim, R. (1978). Obtaining information from non-random missing data. Proceedings of the Section on Statistical Computing, American Statistical Association, 1978, 34-39.

O'Neil, M.J. (1979). Estimating the nonresponse bias due to refusals in telephone surveys. Public Opinion Quarterly, 43, 218-232.

Oh, H.L. and Scheuren, F. (1978a). Multivariate raking ratio estimation in the 1973 Exact Match Study. Imputation and Editing of Faulty or Missing Survey Data, U.S. Department of Commerce, 120-127. Proceedings of the Section on Survey Research Methods, American Statistical Association, 1978, 716-722.

Oh, H.L. and Scheuren, F. (1978b). Some unresolved application issues in raking ratio estimation. Imputation and Editing of Faulty or Missing Survey Data, U.S. Department of Commerce, 128-135. Proceedings of the Section on Survey Research Methods, American Statistical Association, 1978, 723-728.

Oh, H.L. and Scheuren, F. (1980). Estimating the variance impact of missing CPS income data. Proceedings of the Section on Survey Research Methods, American Statistical Association, 1980, 408-415.

Oh, H.L., Scheuren, F. and Nisselson, H. (1980). Differential bias impacts of alternative Census Bureau hot deck procedures for imputing missing CPS income data. Proceedings of the Section on Survey Research Methods, American Statistical Association, 1980, 416-420.

Orchard, T. and Woodbury, M.A. (1972). A missing information principle: theory and applications. Proceedings of the 6th Berkeley Symposium on Mathematical Statistics and Probability, 1, 697-715.

Palmer, S. (1967). On the character and influence of nonresponse in the Current Population Survey. Proceedings of the Social Statistics Section, American Statistical Association, 1967, 73-80.

Palmer, S. and Jones, C. (1966). A look at alternate imputation procedures for CPS noninterviews. U.S. Bureau of the Census memorandum, Washington, D.C.

Patrick, C.A. (1978). Estimation, imputation, randomization, and risk equivalence. Imputation and Editing of Faulty or Missing Survey Data, U.S. Department of Commerce, 94-99. Proceedings of the Section on Survey Research Methods, American Statistical Association, 1978, 486-491.

Platek, R. (1977). Some factors affecting nonresponse. Bulletin of the International Statistical Institute, 47(3), 347-366. Also, Survey Methodology, 3, 191-214.

Platek, R. (1978). Imputation for household surveys in Statistics Canada. European Statistical Conference, Geneva, March 1978.

Platek, R. (1980). Causes of incomplete data, adjustments and effects. Survey Methodology, 6, 93-132.

Platek, R. and Gray, G.B. (1978). Nonresponse and imputation. Survey Methodology, 4, 144-177.

Platek, R. and Gray, G.B. (1979). Methodology and application of adjustments for nonresponse. Bulletin of the International Statistical Institute, 48.

Platek, R., Singh, M.P. and Tremblay, V. (1978). Adjustment for nonresponse in surveys. Survey Sampling and Measurement (Namboodiri, N.K. ed.), Chapter 11. Academic Press, New York.

Politz, A. and Simmons, W. (1949). I. An attempt to get the 'not at homes' into the sample without callbacks. II. Further theoretical considerations regarding the plan for eliminating callbacks. Journal of the American Statistical Association, 44, 9-31.

Politz, A. and Simmons, W. (1950). Note on an attempt to get the 'not at homes' into the sample without callbacks. Journal of the American Statistical Association, 45, 136-137.

Preece, D. A. (1971). Iterative procedures for missing values in experiments. Technometrics, 13, 743-753.

Pregibon, D. (1977). Typical survey data: estimation and imputation. Survey Methodology, 2, 70-102.

Pritzker, L., Ogus, J. and Hansen, M.H. (1966). Computer editing methods - some applications and results. Bulletin of the International Statistical Institute, 41(1), 442-472.

Proctor, C. H. (1978). More on imputing versus deleting when estimating scale scores. Imputation and Editing of Faulty or Missing Survey Data, U.S. Department of Commerce, 36-43. Proceedings of the Section on Survey Research Methods, American Statistical Association, 1978, 209-211.

Purcell, N.J. and Kish, L. (1980). Postcensal estimates for local areas (or domains). International Statistical Review, 48, 3-18.

Radner, D. (1978). The development of statistical matching in economics. Imputation and Editing of Faulty or Missing Survey Data, U.S. Department of Commerce, 108-113. Proceedings of the Social Statistics Section, American Statistical Association, 1978, 503-508.

Raj, D. (1968). Sampling Theory. McGraw-Hill, New York.

Rao, J.N.K. and Hughes, E. (1979). Comparison of domains in the presence of nonresponse. Symposium on Incomplete Data: Preliminary Proceedings (Panel on Incomplete Data of the Committee on National Statistics/National Research Council), 302-316. U.S. Department of Health, Education, and Welfare, Washington, D.C.

Rao, P.S.R.S. (1979). Call-backs, follow-ups, and repeated telephone calls. Draft paper for Panel on Incomplete Data, Committee on National Statistics, National Academy of Sciences.

Rao, P.S.R.S. (1979). Nonresponse and double sampling. Draft paper for Panel on Incomplete Data, Committee on National Statistics, National Academy of Sciences.

Robison, E.L. and Richardson, W.J. (1978). Editing and imputation of the 1977 Truck Inventory and Use Survey. Imputation and Editing of Faulty or Missing Survey Data, U.S. Department of Commerce, 30-35. Proceedings of the Section on Survey Research Methods, American Statistical Association, 1978, 203-208.

Rockwell, R.C. (1975). An investigation of imputation and differential quality of data in the 1970 Census. Journal of the American Statistical Association, 70, 39-42.

Roshwalb, A. (1979). Financial incentives. Draft paper for Panel on Incomplete Data, Committee on National Statistics, National Academy of Sciences.

Roshwalb, A. (1979). Respondent rules. Draft paper for Panel on Incomplete Data, Committee on National Statistics, National Academy of Sciences.

Roshwalb, I. (1953). Effect of weighting by card-duplication on the efficiency of survey results. Journal of the American Statistical Association, 48, 773-777.

Rubin, D.B. (1974). Characterizing the estimation of parameters in incomplete-data problems. Journal of the American Statistical Association, 69, 467-474.

Rubin, D.B. (1976). Inference and missing data. Biometrika, 63, 581-592.

Rubin, D.B. (1976). Comparing regressions when some predictor values are missing. Technometrics, 18, 201-205.

Rubin, D.B. (1977a). Formalizing subjective notions about the effect of nonrespondents in sample surveys. Journal of the American Statistical Association, 72, 538-543.

Rubin, D.B. (1977b). The design of a general and flexible system for handling nonresponse in sample surveys. Unpublished, Working Paper, Educational Testing Service.

Rubin, D.B. (1978). Multiple imputations in sample surveys: a phenomenological Bayesian approach to nonresponse. Imputation and Editing of Faulty or Missing Survey Data, U.S. Department of Commerce, 1-22. Proceedings of the Section on Survey Research Methods, American Statistical Association, 1978, 20-34.

Rubin, D.B. (1979a). Conceptual issues in the presence of nonresponse. Draft paper for Panel on Incomplete Data, Committee on National Statistics, National Academy of Sciences.

Rubin, D.B. (1979b). Illustrating the use of multiple imputations to handle nonresponse in sample surveys. Bulletin of the International Statistical Institute, 1979.

Rubin, D.B. (1980). Handling nonresponse in sample surveys by multiple imputations. U. S. Bureau of the Census, Washington, D.C.

Rust, K. (undated). Nonresponse adjustment in ABS surveys. Australian Bureau of Statistics, Canberra.

Sande, G. (1979a). Replacement for a ten minute gap. Symposium on Incomplete Data; Preliminary Proceedings (Panel on Incomplete Data of the Committee on National Statistics/National Research Council), 481-483. U.S. Department of Health, Education, and Welfare, Washington, D.C.

Sande, G. (1979b). Numerical edit and imputation. International Association for Statistical Computing, 42nd Session of International Statistical Institute, 1979.

Sande, I.G. (1979a). A personal view of hot deck imputation procedures. Survey Methodology, 5, 238-258.

Sande, I.G. (1979b). Hot deck imputation procedures. Symposium on Incomplete Data: Preliminary Proceedings (Panel on Incomplete Data of the Committee on National Statistics/National Research Council), 484-498. U.S. Department of Health, Education, and Welfare, Washington, D.C.

Sande, I.G. (1981). Imputation in surveys: coping with reality. Survey Methodology, 7, 21-43.

Santos, R.L. (1981a). Effects of Imputation on Complex Statistics, Survey Research Center, University of Michigan, Ann Arbor.

Santos, R.L. (1981b). Effects of imputation on regression coefficients. Proceedings of the Section on Survey Research Methods, American Statistical Association, 1981, 140 - 145.

Sarndal, C.E. and Hui, T.-K. (1981). Estimation for nonresponse situations: to what extent must we rely on models? Current Topics in Survey Sampling (D. Krewski, R. Platek and J.N.K. Rao, eds.) pp. 227-246. Academic Press, New York.

Schaible, W.L. (1979). Estimation of finite population totals from incomplete sample data: prediction approach. Symposium on Incomplete Data: Preliminary Proceedings (Panel on Incomplete Data of the Committee on National Statistics/National Research Council), 170-187. U.S. Department of Health, Education, and Welfare, Washington, D.C.

Schaul, R.A. and Hayya, J.C. (1976). An imputation procedure for determining missing factor levels in analysis of variance. Proceedings of the Social Statistics Section, American Statistical Association, 1976, 746-750.

Scheuren, F. (1979). Weighting adjustments for unit nonresponse. Draft paper for Panel on Incomplete Data, Committee on National Statistics, National Academy of Sciences.

Schieber, S.J. (1978). A comparison of three alternative techniques for allocating unreported Social Security Income on the Survey of the Low-Income Aged and Disabled. Imputation and Editing of Faulty or Missing Survey Data, U.S. Department of Commerce, 44-50. Proceedings of the Section on Survey Research Methods, American Statistical Association, 1978, 212-218.

Schore, J. (1980). An analysis of nonresponse to Job Corps evaluation interviews. Proceedings of the Section on Survey Research Methods, American Statistical Association, 1980, 644-648.

Sedransk, J.H. and Singh, B. (1979). Bayesian procedures for survey design when there is nonresponse. Symposium on Incomplete Data: Preliminary Proceedings (Panel on Incomplete Data of the Committee on National Statistics/National Research Council), 317-361. U.S. Department of Health, Education, and Welfare, Washington, D.C.

Shimizu, I.M., Gonzalez, J.F. and Jones, G.K. (1980). Alternative adjustments for nonresponse in the National Hospital Discharge Survey. Proceedings of the Section on Survey Research Methods, American Statistical Association, 1980, 649-651.

Sims, C. (1974). Comment. Annals of Economic and Social Measurement, 2, 395-398.

Singh, B. (1979). Nonresponse and double sampling - the Bayesian approach. Draft paper for Panel on Incomplete Data, Committee on National Statistics, National Academy of Sciences.

Singh, B. and Sedransk, J. (1975). Estimation of regression coefficients when there is nonresponse. Technical Report 25, Statistical Science Division, SUNY, Buffalo.

Singh, B. and Sedransk, J. (1978). Sample size selection in regression analysis when there is nonresponse. Journal of the American Statistical Association, 73, 362-365.

Sirken, M. (1979). Handling missing data using network sampling. Draft paper for Panel on Incomplete Data, Committee on National Statistics, National Academy of Sciences.

Smouse, E.P. (1982). Bayesian estimation of a finite population total using auxiliary information in the presence of nonresponse. Journal of the American Statistical Association, 77, 97 - 102.

Sonquist, J.A., Baker, E.L. and Morgan, J.N. (1974, rev. ed.). Searching for Structure. Institute for Social Research, University of Michigan, Ann Arbor.

Spiers, E.F. and Knott, J.J. (1969). Computer method to process missing income and work experience information in the current population survey. Proceedings of the Social Statistics Section, American Statistical Association, 1969, 289-297.

Statistical Analysis Group in Education (SAGE) (1980). Guidebook for Imputation of Missing Data. American Institutes for Research, Palo Alto, California.

Stemp, P. (1976). Nonresponse in the Labour Force Survey and methods of imputation. Unpublished paper, Australian Bureau of Statistics.

Stephan, F.F. (1945). The expected value and variance of the reciprocal and other negative powers of a positive Bernouillian variate. Annals of Mathematical Statistics, 16, 50-61.

Stopher, P.R. and Sheskin, I.M. (1981). A method for determining and reducing nonresponse bias. Proceedings of the Section on Survey Research Methods, American Statistical Association, 1981, 252 - 257.

Szameitat, K. and Zindler, H. (1965). The reduction of errors in statistics by automatic corrections. Bulletin of the International Statistical Institute Proceedings, 35th Session, 16, 395-417.

Taffel, S., Johnson, D. and Heuser, R. (1982). A method of imputing length of gestation on birth certificates. Vital and Health Statistics, Series 2, No. 93. U.S. Government Printing Office, Washington, D.C.

Thomsen, I. (1973). A note on the efficiency of weighting subclass means to reduce the effects of nonresponse when analyzing survey data. Statistisk Tidskrift, 4, 278-283.

Thomsen, I. and Siring, E. (1979). On the causes and effects of nonresponse: Norwegian experiences. Symposium on Incomplete Data: Preliminary Proceedings (Panel on Incomplete Data of the Committee on National Statistics/National Research Council), 21-62. U.S. Department of Health, Education, and Welfare, Washington, D.C.

Thornberry Jr., O.T. and Massey, J.T. (1978). Correcting for undercoverage bias in random digit dialed national health surveys. Imputation and Editing of Faulty or Missing Survey Data, U.S. Department of Commerce, 56-61. Proceedings of the Section on Survey Research Methods, American Statistical Association, 1978, 224-229.

Timm, N.H. (1970). The estimation of variance-covariance and correlation matrices from incomplete data. Psychometrika, 35, 417-438.

Trawinski, I.M. and Bargmann, R.F. (1964). Maximum likelihood estimation with incomplete multivariate data. Annals of Mathematical Statistics, 35, 647-657.

Trewin, D. (1977). The use of post-stratification for adjustment of nonresponse biases. Bulletin of the International Statistical Institute, 47 (4), 708-712.

Tupek, A.R. and Richardson, W.J. (1978). Use of ratio estimates to compensate for nonresponse bias in certain economic surveys. Imputation and Editing of Faulty or Missing Survey Data, U.S. Department of Commerce, 24-29. Proceedings of the Section on Survey Research Methods, American Statistical Association, 1978, 197-202.

Vacek, P.M. and Ashikaga, T. (1980). An examination of the nearest neighbor rule for imputing missing values. Proceedings of the Statistical Computing Section, American Statistical Association, 1980, 326-331.

Van Guilder, M. and Azen, S.P. (1981). Conclusions regarding algorithms for handling incomplete data. Proceedings of the Section on Statistical Computing, American Statistical Association, 1981.

Vaughan, D.R. (1978). Errors in reporting Supplemental Security Income recipiency in a pilot household survey. Imputation and Editing of Faulty or Missing Survey Data, U.S. Department of Commerce, 142-148. Proceedings of the Section on Survey Research Methods, American Statistical Association, 1978, 288-293.

Welniak, E.J. and Coder, J.F. (1980). A measure of the bias in the March CPS earnings imputation system. Proceedings of the Section on Survey Research Methods, American Statistical Association, 1980, 421-425.

Wilcox, J.B. (1977). The interaction of refusal and not-at-home bias. Journal of Marketing Research, 14, 592-597.

Wilkinson, G.N. (1958). Estimation of the missing values for the analysis of incomplete data. Biometrics, 14, 257-286.

Wilks, S.S. (1932). Moments and distributions of estimates of population parameters from fragmentary samples. Annals of Mathematical Statistics, 3, 163-195.

Williams, R.L. and Folsom, R.E. (1981). Weighted hot-deck imputation of medical expenditures based on a record check subsample. Proceedings of the Section on Survey Research Methods, American Statistical Association, 1981, 406 - 411.

Williams, W.H. and Mallows, C.L. (1970). Systematic biases in panel surveys due to differential nonresponse. Journal of the American Statistical Association, 65, 1338-1349.

Woodbury, M.A. (1979). Statistical record matching for files. Symposium on Incomplete Data: Preliminary Proceedings (Panel on Incomplete Data of the Committee on National Statistics/National Research Council), 236-250. U.S. Department of Health, Education, and Welfare, Washington, D.C.

Woodbury, M.A. and Hasselblad, V. (1970). Maximum likelihood estimation of the variance covariance from the multivariate normal. Proceedings of SHARE, 34, 1550-1561.

Woodbury, M.A. and Siler, W. (1966). Factor analysis with missing data. Annals of the New York Academy of Science, 128, 746-754.

Woolson, R.F. and Cole, J.W.L. (1974). Comparing means of correlated variates with missing data. Communications in Statistics, 3(10), 941-948.

Woolson, R.F., Leeper, J.D. and Clarke, W.R. (1978). Analysis of incomplete data from longitudinal and mixed longitudinal studies. Journal of the Royal Statistical Society, A, 141, 242-252.

World Fertility Survey (1979). Guidelines for Data Processing. World Fertility Survey, London.

Yates, F. (1933). The analysis of replicated experiments when the field results are incomplete. The Empire Journal of Experimental Agriculture, 1, 129-142.

Ycas, M.A. and Lininger, C.A. (1980). The Income Survey Development Program: a review. Proceedings of the Section on Survey Research Methods, American Statistical Association, 1980, 486-490.

ISR RESEARCH REPORTS

The following Research Reports have been published by ISR. They are available in paperbound editions only. For information on prices and availability, write to the ISR Publishing Division, P.O. Box 1248, Ann Arbor, Michigan 48106.

Residential Displacement in the U.S., 1970-1977. Sandra J. Newman and Michael S. Owen. 1982. 98 pp.

Sex Role Attitudes among High School Seniors: Views about Work and Family Life. A Regula Herzog and Jerald G. Bachman. 1982. 272 pp.

Subjective Well-Being among Different Age Groups. A Regula Herzog, Willard L. Rodgers, and Joseph Woodworth. 1982. 115 pp.

Employee Ownership. Michael Conte, Arnold S. Tannenbaum, and Donna McCulloch. 1981. 70 pp.

Recreation and Quality of Urban Life: Recreational Resources, Behaviors, and Evaluations of People in the Detroit Region. Robert W. Marans and J. Mark Fly. 1981. 240 pp.

A Comparative Study of the Organization and Performance of Hospital Emergency Services. Basil S. Grorgopoulos and Robert A. Cooke. 1980. 512 pp.

An Evaluation of "Freestyle": A Television Series to Reduce Sex-Role Stereotypes. Jerome Johnston, James Ettema, and Terrence Davidson. 1980. 308 pp.

Occupational Stress and the Mental and Physical Health of Factory Workers. James S. House. 1980. 356 pp.

Job Demands and Worker Health: Main Effects and Occupational Differences. Robert D. Caplan, Sidney Cobb, John R. P. French, Jr., R. Van Harrison, and S. R. Pinneau, Jr. 1980. 342 pp.

Perceptions of Life Quality in Rural America: An Analysis of Survey Data from Four Studies. Robert W. Marans and Donald A. Dillman, with the assistance of Janet Keller. 1980. 118 pp.

Social Support and Patient Adherence: Experimental and Survey Findings. Robert D. Caplan, R. Van Harrison, Retha V. Wellons, and John R. P. French, Jr. 1980. 283 pp.

Working Together: A Study of Cooperation among Producers, Educators, and Researchers to Create Educational Television. James S. Ettema. 1980. 220 pp.

Experiments in Interviewing Techniques: Field Experiments in Health Reporting, 1971-1977. Edited by Charles F. Cannell, Lois Oksenberg, and Jean M. Converse. 1979. 446 pp.

The 1977 Quality of Employment Survey: Descriptive Statistics, with Comparison Data from the 1969-70 and 1972-73 Surveys. Robert P. Quinn and Graham L. Staines. 1979. 364 pp.

The Physical Environment and the Learning Process: A Survey of Recent Research. Jonathan King, Robert W. Marans, and associates. 1979. 92 pp.

Results of Two National Surveys of Philanthropic Activity. James N. Morgan, Richard F. Dye, and Judith H. Hybels. 1979. 204 pp.

A Survey of American Gambling Attitudes and Behavior. Maureen Kallick, Daniel Suits, Ted Dielman, and Judith Hybels. 1979. 560 pp.